Ecological Studies, Vol. 106

Analysis and Synthesis

Edited by

O.L. Lange, Würzburg, FRG
H.A. Mooney, Stanford, USA
H. Remmert, Marburg, FRG

Ecological Studies

Volumes published since 1988 are listed at the end of this book

Hermann Remmert (Ed.)

Minimum Animal Populations

With 75 Figures, Some in Colour

Springer-Verlag
Berlin Heidelberg New York
London Paris Tokyo
Hong Kong Barcelona
Budapest

Professor Dr. Hermann Remmert
Universität Marburg
Fachbereich Biologie-Zoologie
D-35032 Marburg
Germany

ISBN 3-540-56684-8 Springer-Verlag Berlin Heidelberg New York
ISBN 0-387-56684-8 Springer-Verlag New York Berlin Heidelberg

Library of Congress Cataloging-in-Publication Data. Minimum animal populations / Hermann Remmert (ed.). p. cm.—(Ecological studies; vol. 106) Includes bibliographical references and index. ISBN 3-540-56684-8 (Berlin: acid-free paper).—ISBN 0-387-56684-8 (New York: acid-free paper) 1. Animal populations. 2. Extinction (Biology) I. Remmert, Hermann. II. Series: Ecological studies; v. 106. QL 752.M56 1994 591.52′48—dc20 93-340 10

Printed in Germany

Typesetting: Thomson Press (India) Ltd., New Delhi
31/3145/SPS - 5 4 3 2 1 0 - Printed on acid-free paper

Contents

1 Introduction
H. Remmert 1

2 The Viability of Small Populations of Birds: an Empirical Investigation of Vulnerability
P.G. Ryan and W.R. Siegfried 3

2.1 Introduction 3
2.2 The Theory of Small Population Extinctions 4
2.3 Evidence from Birds: Genetic Effects 7
2.4 Persistence of Small Populations 10
2.5 Conclusions 17
References 18

3 Small World Populations in Birds: An Attempt of a Brief General Survey
E. Bezzel 23

3.1 Introduction 23
3.2 Material and Methods 24
3.3 Results and Conclusions 27
References 31

4 Genetic, Demographic, Spatial, Environmental and Catastrophic Effects on the Survival Probability of Small Populations of Mammals
H. Korn 33

4.1 Introduction 33
4.2 Genetic Considerations 33
4.3 Demographic and Spatial Considerations 39
4.4 Environmental and Catastrophic Considerations 44
4.5 Conclusions 45
References 47

5 On Micropopulations in Fish and Their Effects on Differentiation and Speciation
W. Villwock 51

5.1 Introduction 51
5.2 Micropopulations in Anatolian *Aphanius* ssp. (Fam.: Cyprinodontidae) 51
5.3 Intralacustrine Segregation Phenomena in East African Cichlids (Cichlidae) 55
5.4 Conclusion 64
References 64

6 Modelling Extinction and Survival of Small Populations
C. Wissel, T. Stephan, and S.-H. Zaschke 67

6.1 Introduction 67
6.2 Population Dynamics with Overlapping Generations . . 68
6.3 Population Dynamics with Non-Overlapping Generations. 83
6.4 Conclusion 95
References 103

7 "Very Small" Bird Populations in Amazonia
J. Haffer 105

7.1 Introduction 105
7.2 "Very Small" Bird Populations in Amazonia 105
7.3 Summary 115
References 116

8 Minimum Population of *Clethrionomys glareolus* (Schreber 1780)
G. Bujalska 119

8.1 Introduction 119
8.2 Area, Material and Methods 119
8.3 Results 124
8.4 Discussion 128
References 129

9 Minimum Populations of Carabid Beetles (Col., Carabidae)
L. Grüm 131

9.1 Introduction 131
9.2 Area and Methods 131
9.3 Results 132
9.4 Discussion 135
References 136

10 To Survive or To Become Extinct: Small Populations of Tetraonids in Central Europe
S. Klaus 137

10.1 Introduction 137
10.2 Capercaillie 139
10.3 Black Grouse 142
10.4 Hazel Grouse 150
10.5 Conclusions 150
References 151

Subject Index 153

1 Introduction

H. Remmert

Small populations are very often discussed, but there seems to be no general overview touching all the self-evident but normally simply neglected problems connected with small populations. First, there are many very different types of organisms, and as every biologist should know, the problems of small populations are very different in different types of organisms.

1. In vascular plants the problems are different from the situation in birds and mammals; in marine benthic animals or in parasites such as tapeworms the problems are different again, and in seasonal planktonic animals or insects they are different from those in biotopes under constant conditions. In tapeworms or in vascular plants, an adult organism seems to be comparable to a population of mammals or birds because its offspring can be so diverse and plentiful.

2. There are small populations which explode and break down to a small population again, and then explode and break down again.

3. The third possibility is the long-term existence of very small populations—on oceanic islands or ecological islands—isolated from the main population or without such a main population. Although very few, there seem to be such examples. The thrushes specific to the Tristan da Cunha group seem to exist in populations of less than 100 pairs; the bearded vulture seems to exist, apart from its main range in central Asia, in very small populations on high mountains in Southern and Eastern Africa (about 20 pairs in South Africa) in very small populations on Sardinia, Corsika, Rhodos and Kreta, and probably in not much greater populations in Spain and the former Yugoslawia. An obviously isolated, very small population of Andean condors lives in the Argentinian Sierra de Cordoba, and introgression is not discernible. Furthermore, very small populations in African mammals seem to exist: the colonies, e.g. of the naked mole rat in East Africa, seem to exist without any intercolonial contact and, after Rasa the same seems to hold for the colonies of the dwarf mungo, although in the latter case there are different opinions.

In such cases of very small populations, the ecologist usually asks first after the genetical influence of inbreeding; but, looking at the examples, the dangerous effects of inbreeding appear to play a minor role. There seems to be no inbreeding depression in naked mole rats or bearded vultures. The fact that quite a number of vertebrate populations exist in the field without any genetical differences between the individuals or even between the populations (as is well known in naked mole rats and cheetas) makes genetic problems due to inbreeding very improbable. However, all our golden hamsters in laboratories, zoos or pet animals descend from one pregnant female from Syria; similar experiences exist all over the world with naturalized mammals and birds (Long 1981; Lever 1984). The muskrats throughout Europe descend from very few specimens released in the former Czechosowakia, all the racoons in Europe from very few specimens released in Hessen/Germany; a great part of the wild turkeys now existing in the United States descend from a few individuals released in the Black Hills after their extermination in this same area. The two books mentioned give a great number of such examples, so we have to conclude that deleterious effects of inbreeding are not the rule, but rather the exception.

Of course, very small populations can be exterminated by unforseeable chance effects such as outbreaks of volcances or the like. Such very small populations are therefore more endangered than large populations, but inbreeding can be neglected.

Rather, we can assume that in natural biotopes with their mosaic-cycle process, organisms continuously fluctuate in population size, and continuously have to found new populations, starting with few founding specimens. This is only possible if the new populations increase rapidly and send out new founding members to new places which come into existence during the mosaic-cycle process. Small animals and r-strategists are more involved in this than large organisms or K-strategists. The effect is always the same: very many populations have to go through very small numbers of individuals (bottlenecks) and have to recover from these very small numbers. They would no longer exist if they were not able to survive such a situation.

References

Lever C (1985) Naturalized mammals of the world. Longman, London

Long JL (1981) Introduced birds of the world. David & Charles, Newton, Abbot, London

Remmert H (ed) (1991) The mosaic-cycle concept of ecosystems. Ecological Studies, vol 85. Springer, Berlin Heidelberg New York

2 The Viability of Small Populations of Birds: an Empirical Investigation of Vulnerability

P.G. RYAN and W.R. SIEGFRIED

2.1 Introduction

The likelihood of the extinction of a significant portion of the world's biota during the next several decades is cause for concern. This has given rise to the discipline of conservation biology, with the concomitant attempt to predict minimum viable population sizes (MVP) as one of its central themes. A substantial literature on MVP has arisen in the decade since the concept was introduced (Shaffer 1981; Schonewald-Cox et al. 1983; Soulé 1986, 1987a; Simberloff 1988), but guidelines for estimating MVP have been based primarily on theoretical considerations (Simberloff 1988; Thomas 1990). There are good reasons for this bias.

Apart from the prohibitive scale necessary for experimental investigation of MVP (Simberloff 1988), extinction is a probabilistic event at our current level of understanding (e.g. Shaffer 1987), confounding attempts to predict the behaviour of one population on the known performance of a similar population. A case in point is the extinction of two house sparrow *Passer domesticus* populations introduced to the east coast of North America prior to the species' successful introduction in 1853 (Long 1981). Moreover, models estimating MVP are far from perfect (Harris et al. 1987; Soulé 1987b). They are based on numerous assmptions which, while describing general trends, may be misleading when applied to individual cases. For example, MVP model predictions for a particular species ignore interactions with other species which can play important roles in the maintenance or extinction of populations (e.g. Mayfield 1983; Lovejoy et al. 1984; Zimmerman and Bierregaard 1986). Hence, it is now generally accepted that only a detailed investigation of all the factors affecting a population (population viability analysis, PVA) can assess the risk of extinction (Gilpin and Soulé 1986; Shaffer 1990), although even PVA fails to incorporate unprecedented events which, by definition, cannot be predicted (Simberloff 1988). However, for most threatened species, a suitable database is lacking for undertaking PVA, and there is neither the time nor the support available for such data to be collected (Gilpin 1987; Shaffer 1987). Consequently, there is a need to provide nature-conservation managers with broad guidelines to estimate MVP for any given population.

Studies of the dynamics of "real" populations (i.e. specific case histories of populations) can prove to be valuable, by allowing an independent examination of the reliability of MVP guidelines derived from population theory (Pimm et al. 1988; Thomas 1990). This is particularly important given the large population sizes predicted by some models to be necessary to have a reasonable chance of a population with evolutionary potential surviving over even a modest time span (e.g. Leigh 1981; Belovsky 1987; Goodman 1987a). The predictions of such models have been used to justify disregarding the protection of very small populations or closing small nature reserves (Simberloff 1988), despite Soulé's (1987b) assertion that there are no hopeless cases.

Here, we report on an empirical examination of the vulnerability of small populations of birds. The focus on birds has two advantages: birds are one of the best-studied groups of animals, and, as relatively large organisms, they can be useful target species for the conservation of ecosystems. It is emphasized that the purpose of this chapter is in no way prescriptive: we do not attempt to provide revised estimates of MVP. The aim is rather to draw attention to some possible inadequacies of current MVP models and thus provide a more balanced perspective potentially useful to nature-conservation managers who are obliged to make decisions regarding the fate of small populations in the absence of adequate information.

2.2 The Theory of Small Population Extinctions

The factors which combine to enhance the risk of extinction in small populations have been extensively reviewed (e.g. Gilpin and Soulé 1986; Soulé 1987a; Simberloff 1988). We shall only briefly consider the main points here. Three types of chance events influence the likelihood of extinction: genetic, demographic and environmental stochasticity (Shaffer 1981, 1987). However, due to the frequent confusion between demographic and environmental factors (Simberloff 1988) and the tendency to model genetic effects independently from demographic/environmental effects (Ewens et al. 1987), it is simpler to consider two levels of stochastic events: those operating at the level of the gene pool (genetic effects), and those at the level of the individual (demographic and environmental effects). It must be borne in mind, however, that these factors act in concert (the so-called extinction vortex) to increase the likelihood of extinction of small populations (Gilpin and Soulé 1986).

2.2.1 Genetic Stochasticity

In sexually reproducing populations, genetic variation is gradually reduced by chance events (drift, through sampling error) in the absence of mutation, migration and selection. This loss of genetic diversity reduces the potential for evolutionary responses to changing environmental conditions, and may lower fitness through the loss of heterozygosity. Models estimating the rate of loss of quantitative genetic variation suggest that, at equilibrium, an effective population size N_e of the order of 500 is sufficient for mutation to counter drift, even under conditions of stabilizing or fluctuating directional selection (Lande and Barrowclough 1987; but see Lande 1988 for a discussion of the limitations of this number). A much larger population size (of the order of 10^5) is required to counter the loss of neutral alleles, but this probably has less evolutionary significance than variability in quantitative traits (Simberloff 1988). There is considerable loss of genetic diversity when population size is temporarily very small (a "bottleneck"), particularly when recovery is slow or bottlenecks are repeated (Motro and Thomson 1982; Lande and Barrowclough 1987). This should favour species with faster rates of reproduction r, which can rapidly increase population size after a bottleneck. Founder events, when a small number of individuals colonizes a new area, form a subset of population bottlenecks.

A related threat facing small populations at the genetic level is inbreeding depression. This occurs when genetically similar individuals interbreed, either as a result of small population size or breeding with close relatives, and results from the expression of deleterious alleles (Lande and Barrowclough 1987). In small populations, some deleterious alleles can become fixed by chance if they are not lethal when homozygous, reducing the fitness of the entire population. However, a population can adapt to high levels of inbreeding, especially if population size decreases gradually (Templeton and Read 1983; Simberloff 1988).

The inbreeding coefficient (F) increases by $1/2N_e$ per generation and, based on observations by animal breeders that increases in $F > 1\%$ per generation are unacceptable, it has been suggested that inbreeding is a problem only when N_e is 50 or less (e.g. Franklin 1980; Soulé 1980). It should be noted that the calculation of inbreeding N_e differs from variance N_e used to estimate the effects of genetic drift (see Crow and Kimura 1970). However, the importance of inbreeding depression on recruitment is unresolved in species which produce several offspring per breeding attempt. Given density-dependent mortality of immatures, inbreeding depression may have little influence on the numbers of recruits to the breeding population, because most adverse effects arise during the early stages of development (Shields 1982; Greenwood 1987; but see Charlesworth and Charlesworth 1987).

Estimation of N_e requires detailed demographic data, including the type of breeding sytem, numbers of breeding males and females, and variability in lifetime reproductive success (Lande and Barrowclough 1987). For birds, N_e for any given generation is generally less than half the total population size N, as a result of the large proportion of non-breeding birds and the considerable interindividual variation in lifetime reproductive success (e.g. Newton 1989; Grant and Grant 1989). When population size fluctuates, long-term N_e is approximated by the harmonic mean of successive generations' N_e (Crow and Kimura 1970). This results in small population sizes having a disproportionately large influence on long-term N_e, which is consistent with extensive drift occurring during bottlenecks and founder events.

2.2.2 Demographic and Environmental Stochasticity

At the level of the individual, several factors together influence the probability of a population becoming extinct. In the absence of environmental variation, demographic stochasticity, which results from demographic "accidents" (e.g. whether individuals live or die, and the number and sex ratio of their offspring), rarely results in extinction unless the population size is very small (40 or less, depending on the population growth rate r). This is because the time to extinction T increases as a function of a constant to the power N_m, the maximum population size (Goodman 1987b; Fig. 2.1). For any given N_m, persistence time, in numbers of generations, increases with increasing r (Leigh 1981). Since longevity and body size are negatively correlated with r, small, fast-reproducing

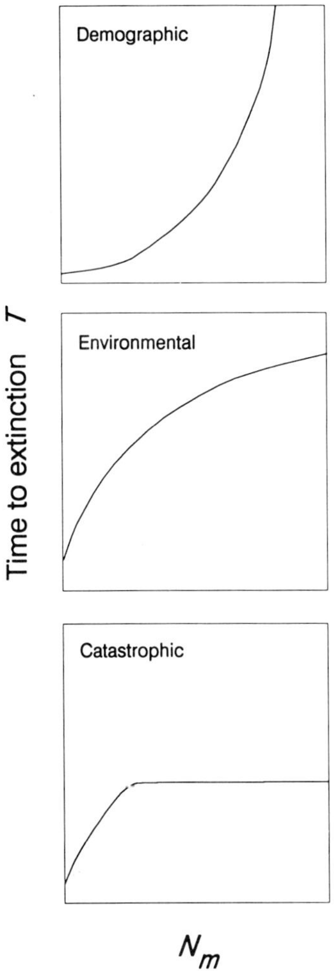

Fig. 2.1. Relationships between T, time to extinction, and N_m, maximum population size, used to model demographic, environmental, and catastrophic stochasticity

organisms tend to persist longer than large, slower-reproducing organisms (Pimm et al. 1988).

Environmental stochasticity encompasses a continuum of disturbances ranging from short-term fluctuations to long-term variation and catastrophes. Shaffer (1981, 1987) considered catastrophes to be distinct from environmental stochasticity, but there is no clear division between these categories which are dependent in part on scale (e.g. a heavy downpour of rain to a large organism could be a catastrophic flood to a population of small organisms, cf. Simberloff 1988). The risk of extinction from environmental stochasticity is less sensitive to population size than is the risk resulting from demographic stochasticity, but this varies in relation to the severity of the perturbation. Models of the effect of moderate environmental disturbances on the risk of extinction have assumed the time to extinction T to be a function of the logarithm of N_m (Leigh 1981; Fig. 1), whereas catastrophes have been assumed to cause extinction irrespective of N_m (Pimm et al. 1988; Fig. 2.1). Clearly these models are simplistic, given a continuum of perturbation severity. An alternative approach is to use population variabilities recorded for natural populations over long intervals to determine the approximate limits of environmental stochasticity for model simulations (Pimm and Redfearn 1988; Thomas 1990). Other factors being equal, a larger temporal variance in population size should increase the risk of extinction (Pimm et al. 1988).

2.3 Evidence from Birds: Genetic Effects

Given this brief overview of the factors known to influence the risk of extinction, we can examine the performance of small populations of birds to assess the relative importance of these factors. There are two sources of evidence: controlled experiments and inference from natural populations. Given the relatively long generation times of birds, there have been few controlled studies of their populations. No experiment has directly addressed the risk of extinction of different bird populations, but there have been several experiments involving birds that provide insight into the theoretical background outlined above, particularly as regards genetic effects.

2.3.1 Evolutionary Potential

It is well established that the potential for evolution is restricted by the degree of heritable variability, but how marked is the effect of small population size on this potential? Vasquez and Bohren (1982) showed that among three populations of leghorn chickens, *Gallus domesticus*, with $N_e = 3.6$, 7.3 and 14.6, the response to a constant selection pressure for larger body mass increased as a function of N_e. However, the difference in response was most marked between the two smallest population sizes (Fig. 2.2), suggesting that for $N_e > 10$ there is a considerable scope for an evolutionary response. This result applies to

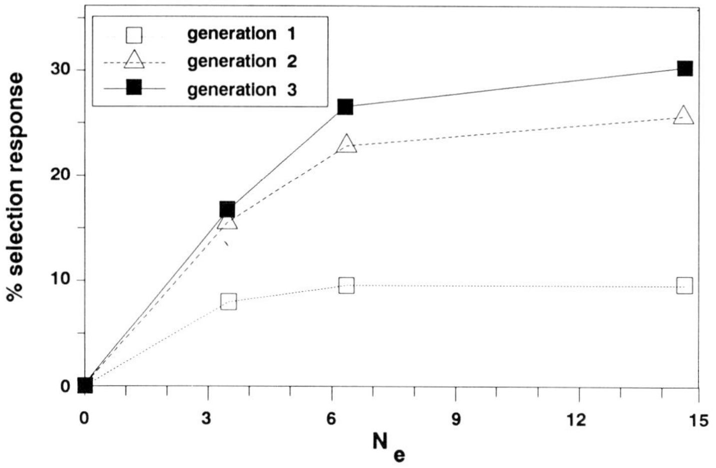

Fig. 2.2. Selection response among successive generations of leghorn chickens as a function of effective population size. (After Vasquez and Bohren 1982)

short-term selection pressures on populations with a large initial population size. There would probably have been much less response if populations had been maintained at a small size for several generations prior to selection; experimental confirmation of this point is required.

In addition to the magnitude of response, the duration of response to selection varied in relation to N_e. All three population sizes of leghorn chickens showed reduced response by the third generation, with the smallest population size showing virtually no response after the second generation (Fig. 2.2). There have been no long-term studies on birds, but those on mice found that only populations with $N_e > 40$ showed a significant response to continuous selection after 14 generations (Eisen 1975). Inferences of the evolutionary potential of small, naturally occurring populations based on their morphological variability are presented below.

2.3.2 Inbreeding Depression

Although not always the case (e.g. Tomlinson et al. 1991), there is experimental evidence that inbreeding depression is an important factor among small captive populations of birds, affecting, inter alia, male fertility, clutch size and egg hatchability (e.g. Sittman et al. 1966; Soulé 1980; Greenwell et al. 1982). These results suggest that inbreeding between close relatives can significantly reduce fitness, and is thus likely to be selected against in natural populations. Close inbreeding (full sibling or parent-sibling, cf. Ralls et al. 1986) may be avoided by kin recognition; Bateson (1982) demonstrated that individuals in a captive population of Japanese quails *Coturnix japonica* mated preferentially with individuals that differed slightly from their parents, discriminating on the basis of plumage differences. Other criteria, such as song structure, could also be used to discriminate closely related kin, even though the criteria used need not be under direct genetic control (Grant and Grant 1989).

How important, however, is inbreeding among natural populations? One line of evidence involves inference from the observed frequency of inbreeding. There have been few studies of inbreeding among natural bird populations (Greenwood 1987; Ralls et al. 1987; Gibbs and Grant 1989), but the low frequency with which close inbreeding typically occurs (generally $<3\%$) has been interpreted as the result of selection against inbreeding (Ralls et al. 1987). Also, Koenig and Pitelka (1979) suggested that there were behavioural mechanisms to reduce close inbreeding among cooperatively breeding Acorn Woodpeckers, *Melanerpes formicivorus*. However, the few studies that have compared the frequency of inbreeding with that expected by random assortment have found no evidence of avoidance (van Noordwijk and Scharloo 1981; Gibbs and Grant 1989).

More direct evidence of the costs of inbreeding can be obtained by comparing the survival and reproductive success of inbred and non-inbred individuals. No differences were found between highly inbred and less inbred splendid wrens, *Malurus splendens* (Rowley et al. 1986), but this species has a relatively high level of inbreeding (Ralls et al. 1986), and is thus likely to be adapted for inbreeding depression (cf. Templeton and Read 1983). Similarly, Craig and Jamieson (1988) found no adverse effects of inbreeding among pukeko, *Porphyrio porphyrio*, with 70% of pairs closely related.

Results vary among species characterized by infrequent inbreeding. Gibbs and Grant (1989) detected no impact of inbreeding on the reproductive success of a population of medium ground finches, *Geospiza fortis*, whereas van Noordwijk and Scharloo (1981) found a marked decrease in egg hatchability with increasing levels of inbreeding among great tits, *Parus major*. This difference may result from the finch population having reduced heterozygosity and fewer deleterious alleles relative to the tit population, as a consequence of greater historical isolation of the finches (Gibbs and Grant 1989; but see below). However, despite the lower hatchability of inbred tit eggs, there was no difference in the number of recruits to the breeding population from inbred and non-inbred tits (van Noordwijk and Scharloo 1981).

2.3.3 Evidence from Island Populations

Island populations typically should have less genetic diversity than mainland populations, because of bottlenecks during colonization (founder effects) and during subsequent population fluctuations (Boag 1988). Very high levels of inbreeding among many island populations (Craig and Jamieson 1988; Craig 1991; but see Boag 1988; Gibbs and Grant 1989) should also reduce heterozygosity. Most island populations of birds conform to this prediction (e.g. Gyllensten et al. 1985; Baker et al. 1990; Fleischer et al., in press), although they are not as homozygous as many island populations of mammals (Kilpatrick 1981; Boag 1988). The greater heterozygosity of bird populations relative to island populations of mammals may be due in part to their greater vagility.

The Galápagos ground finches, *Geospiza* spp., are unusual in having an average heterozygosity equivalent to that of continental taxa (Yang and Patton 1981; Boag 1988). This may result from introgression, the incorporation of genes from other species; interspecific hybridization occurs relatively frequently within *Geospiza* (Grant 1986; Grant and Grant 1989; Grant and Grant 1992). Introgression may be an important source of genetic variation for small populations, increasing their evolutionary potential (e.g. Gill 1980; Cade 1983; Grant and Grant 1989; Grant and Grant 1992). However, there remains the vexing question of loss of genetic identity as a result of excessive introgression. Presumably introgression should be discouraged when it results from secondary contact brought about by human activities (e.g. introductions or range changes resulting from habitat modification).

2.4 Persistence of Small Populations

There is a large literature on the persistence of small populations of birds, based on observations of natural populations (e.g. Willis 1974; Jones and Diamond 1976; Schoener 1976; Terborgh and Winter 1980; Karr 1982a, b; Diamond 1984; Pimm et al. 1988; Soulé et al. 1988, Newmark 1991; Tracy and George 1992). Early studies focussed primarily on species turnover, in the light of island biogeographic theory (MacArthur and Wilson 1967), but recently, greater emphasis has been placed on comparing the attributes of species which persist with those that become extinct, in an attempt to identify factors influencing the risk of extinction.

2.4.1 Population Persistence in Habitat Islands

Studies of population persistence are either direct (i.e. based on long-term observations of population dynamics) or indirect [i.e. infer extinction rates from species loss (= relaxation) in habitat fragments of known age, such as land-bridge islands or man-made habitat islands]. Direct observations allow estimates of actual extinction rates, but the long persistence times of all but the smallest populations restrict analyses to very small populations (Pimm et al. 1988; Tracy and George 1992). Indirect studies are less exact, because assumptions have to be made regarding the size and composition of the pre-isolation community. However, such studies allow examination of the persistence times of larger populations, and may provide further insight into specific attributes that influence the risk of extinction.

There are relatively few long-term data sets for population dynamics of birds that are suitable for measuring extinction rates of known population sizes (Jones and Diamond 1976; Diamond 1984; Pimm et al. 1988). The most extensive data

Table 2.1. Persistence time T of land-bird populations as a function of population size at small islands off the British coast (calculated from Pimm et al. 1988). No extinctions were recorded for populations larger than 18 breeding pairs

Population size (no. of breeding pairs)	Sample size	1/T (years)		
		Mean	SD	Range
<2	122	0.532	0.358	0.0–1.0
2–3.9	76	0.245	0.303	0.0–1.0
4–6.9	53	0.063	0.150	0.0–1.0
7–11.9	35	0.025	0.058	0.0–0.25
12–18	30	0.012	0.038	0.0–0.17

set comes from censuses of breeding land birds at islands off the coast of Britain, which clearly demonstrates that persistence time of bird populations is a function of population size (Diamond 1984; Pimm et al. 1988; Tracy and George 1992; Table 2.1). Populations with a mean size of fewer than ten breeding pairs are likely to become extinct within decades, whereas larger populations seldom become extinct within this time frame (Jones and Diamond 1976, Diamond 1984; Pimm et al. 1988). Several indirect studies have also concluded that rarity (i.e. small initial population size) is the major determinant of extinction during relaxation in habitat fragments (e.g. Terborgh and Winter 1980; Diamond 1984; Bolger et al. 1991; Newmark 1991), but other factors may become more important as population size increases (cf. Karr 1982a, b).

There is considerable residual variation in persistence times once the effects of population size have been removed. In accordance with models of demographic stochasticity, Pimm et al. (1988) suggested that body size (inversely related to r) affected the risk of extinction independently of its effect on population density (and hence population size). Above a threshold of about seven breeding pairs, populations of small birds persiste longer than did those of large birds, whereas in very small populations (<seven pairs), the greater longevity of large birds allowed their populations to persist for longer (in years) than those of small organisms (Pimm et al. 1988; but see Tracy and George 1992). This crossover at small population size presumably would disappear if persistence were expressed in terms of generations.

Karr (1990) suggested that the risk of extinction on Barro Colorado Island, Panama, was greater among species with low adult survival rates. This would appear to counter the relationship between persistence time and r, because adult survival typically is inversely related to r. However, Karr (1990) inferred low reproductive success among birds with low adult survival rates, suggesting that some species are "predisposed" to extinction.

Another factor which may influence the risk of extinction is the degree of temporal variability in population size. Both Karr (1982a) and Pimm et al. (1988) found that variable populations have a greater risk of extinction than those with less variation in population size (but see Tracy and George 1992).

Thomas (1990) has argued from data in Pimm and Redfearn (1988) that variation in population size measured over eight to ten generations gives a good approximation of medium-term levels of variability, but there are insufficient data sets to search for determinants of population variability.

Pimm et al. (1988) also showed that migratory birds had a greater risk of extinction than resident birds, especially on small islands (Tracy and George 1992). This result may not hold for all species, but it appears plausible for many birds. Mayfield (1982) has argued that immature Kirland's warblers, *Dendroica kirtlandii*, fail to recruit because they are less adept at correcting for lateral drift during migration than adults and, thus, often fail to locate the spatially restricted breeding grounds. Small populations dispersed in a patchy environment may experience similar problems finding a mate (e.g. Lande 1987), but the problem is likely to be exacerbated in migratory species.

Finally, some species are prone to extinction because of intrinsic characteristics of the species' biology, such as obligate communal activities (social dysfunction), reliance on other taxa for some aspect of their life cycle (secondary extinctions and cascade effects), or complex habitat requirements and other environmental characteristics. Social dysfunction involves a range of group activities including breeding, hunting and defence against predators (e.g. Soulé 1983). A classic example of secondary extinction follows the loss of army ants from small forest patches in Amazonia, when several species of ant-following birds disappear (Lovejoy et al. 1984). Less is known about the absolute habitat requirements of individual species, but Karr (1982b) suggested that many bird extinctions at Barro Colorado Island were of species that regularly moved between habitats; this was not possible after fragmentation (island formation). These types of risks cannot be assessed without detailed knowledge of community interactions. However, given the low predictive ability of simple models of the risk of extinction (Tracy and George 1992), such factors may be very important.

There is scope for more studies of population persistence times in habitat fragments, which, in addition to the trends already detected, could usefully test for differences between habitats (cf. Diamond 1984; Soulé et al. 1988) and between guilds (cf. Terborgh and Winter 1980; Karr 1982b, Newmark 1991). However, while providing valuable insights into the likely persistence times of small populations in isolated nature reserves, these studies have two drawbacks. The first problem is that it is often impossible to estimate the incidence of immigration during the course of the studies. Recent research (e.g. Payne 1990) has shown that earlier calculations of bird movement and, hence, gene flow (e.g. Barrowclough 1978, 1980), underestimated actual movements by more than an order of magnitude. Immigration would tend to inflate persistence times of small populations by increasing genetic diversity, as well as by boosting population size. Consequently, extinction rates estimated by these techniques must be interpreted as minimum values (Tracy and George 1992).

The second problem is more significant; all the studies for which there are adequate controls (i.e. knowledge of the initial, pre-relaxation community) deal with relatively short time frames (all <100 years). As a result, they give little insight into either the long-term persistence or the evolutionary potential (in terms of genetic diversity) of small populations. Long-time-base studies of relaxation (e.g. Pleistocene land-bridge islands, Wilcox 1980) give estimates of the rate of species loss, but this cannot be related to population size, and it is thus of little use for estimating MVP (Shaffer 1981). Since we cannot afford to wait for long-time series of data to accumulate, we have to use different approaches to investigate the long-term fate of small populations. One approach which has been widely employed is to use simulation models. A second approach involves examining populations that are known to have maintained small population sizes for evolutionary periods (1000+ years).

2.4.2 Performance of Higher Taxa with Small Populations

We can eliminate the problems of movement between populations and examine long-term persistence by considering taxa above the level of the population. To prove useful, such analyses must exclude taxa that have small populations as a result of human activities (either direct or indirect) or other systematic (i.e. deterministic) pressures (Shaffer 1981). This largely restricts the choice of species to islands, where the history of human intervention is documented.

There are few naturally occurring small populations of birds on relatively undisturbed islands. Walter (1990) showed that the endemic subspecies of redtailed hawk, *Buteo jamaicensis socorroensis*, restricted to Socorro Island, eastern Pacific Ocean, has maintained a population of approximately 20 pairs for at least the last 120 years (40 generations). Occasional immigration from continental populations cannot be excluded, but it is unlikely, and gene flow has been sufficiently slow for the island population to develop unique phenotypic characters (Walter 1990).

Immigration can be ruled out in the case of endemic island species. Wilkins' bunting, *Nesospiza wilkinsi*, has two subspecies, one at each of two islands in the Tristan da Cunha island group, central South Atlantic Ocean. Nightingale Island (3 km^2) supports ca. 30 pairs, and Inaccessible Island (13 km^2) fewer than 100 pairs, with marked differences between the two island populations, suggesting very little movement between islands (Collar and Stuart 1985; Ryan unpubl. data). The species feeds extensively on the fruit of *Phylica arborea*, and there is little evidence that habitat availability has altered at the islands during the last 20000 years (Preece et al. 1986).

The mangrove finch *Cactospiza heliobates* is a similar example, found on two islands in the Galápagos Archipelago. The total population of between 100 and 200 birds is restricted to some 500 ha of mangroves, and there is nothing to suggest that the species' range or population size has been substantially reduced in recent times (King 1981). There are several other examples of endemic species that have apparently persisted at very small population sizes for protracted periods (e.g. Weller 1980; King 1981).

How have these species survived for evolutionary periods at population sizes considerably smaller than those predicted by MVP models? It may be that they are simply very lucky; many other species with similarly limited ranges probably have become extinct. Clearly this suggestion is untestable. However, it is useful to question the extent to which these species have been affected by consistently small population sizes. Grant and Grant (1989) suggest that relatively low morphological variability among extinct and near-extinct island populations of finches may have resulted from their small population size, and then contributed to their demise through an inability to respond to changes. The choice of species of perhaps questionable, because their decreases resulted from systematic pressures; the Azores bullfinch *Pyrrhula p. murina*, declined primarily as a result of hunting (King 1981), and the Kona grosbeak, *Chloridops kona*, and greater Koa finch, *Rhodacanthis palmeri*, like all Hawaiian land birds, were severely affected by introductions and habitat destruction (King 1978; Freed et al. 1987). However, the hypothesis is worth testing: do populations with consistently small populations have less morphological variability than those with larger populations?

The Tristan da Cunha group and nearby Gough Island support three bunting species, two of which have two subspecies. There are no significant relationships between population size and morphological variability within this closely related group of birds (Fig. 2.3), although the smallest population has the lowest levels of variability. Relative to other island populations of finches, the degree of variability is small (cf. Grant and Grant 1989), which could result from the loss of genetic diversity among all populations (see above). However, a substantial

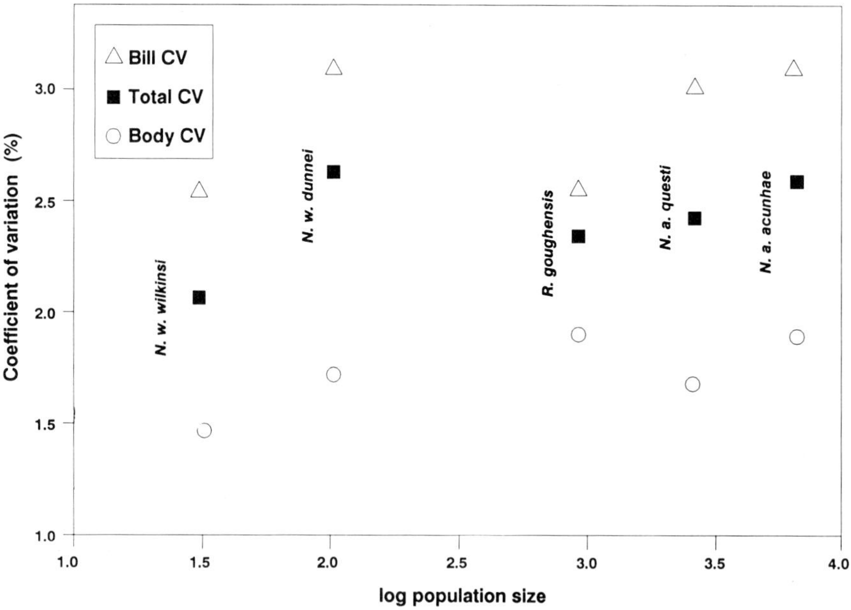

Fig. 2.3. Coefficients of morphological variation as a function of population size (number of breeding pairs) among closely related populations of finches at the Tristan-Gough islands

Table 2.2. Breeding success of Wilkins' bunting *Nesospiza wilkinsi dunnei* (N = 250) and Tristan bunting, *N.a. acunhae* (N = 10 000), at Inaccessible Island during 1989–90 (Ryan, unpubl. data)

Breeding parameter	Wilkins' bunting		Tristan bunting		Significance
Clutch size	1.88	(8)	1.85	(67)	NS
No. hatching	1.50	(8)	1.52	(67)	NS
% Hatching	80.0	(15)	82.2	(124)	NS
No. fledging	0.38	(8)	1.23	(67)	**
% fledging (eggs)	20.0	(15)	66.1	(124)	***
% fledging (chicks)	25.0	(12)	80.4	(102)	***
No. independent[a]	0.38	(8)	0.98	(66)	*
% Independent (eggs)	20.0	(15)	53.3	(122)	*
% Independent (chicks)	25.0	(12)	65.0	(100)	**
% Independent (fledglings)	100.0	(3)	81.3	(80)	NS

[a] Number reaching 50 days old, when juveniles are fully grown and parental support ceases. Sample size are given in parentheses; significance difference between species denoted as follows: *$P < 0.05$, **$P < 0.01$, ***$P < 0.001$ (G and t-tests).

proportion of morphological variability is heritable, at least for the most abundant taxon ($h^2 = 0.48$–0.70; Ryan, unpubl. data), indicating a genetic basis for much of the variability. Inbred lines can be more phenotypically variable than outbred lines (Beardmore 1973; but see Handford 1980), but this is unlikely to be important because the size variation is heritable and the island populations are likely to have adapted to high levels of inbreeding (Craig 1991; see above).

The only adverse effect of persistently small population size detected among Wilkins' bunting was a significantly lower breeding success than its much commoner congener (Table 2.2). Clutch size and hatchability were similar for the two species, but chick mortality was much greater among Wilkins' buntings, as a result of nest desertion and predation by Tristan thrushes, *Nesocichla eremita*. There was no difference in the proportion of fledglings attaining independence, although sample sizes for Wilkins' buntings were small (Table 2.2). It is unclear whether the reduced breeding success results from inbreeding depression and, if so, whether it translates into a smaller number of recruits to the breeding population (cf. van Noordwijk and Scharloo 1981).

The Tristan islands have little seasonal or inter-annual variation in climate (Höflich 1984), which may promote the long-term persistence of small populations cf. Soulé 1987b). By comparison, the Galápagos experience large interannual climatic variation associated with El Niño events, and this results in large fluctuations in the population sizes of finches (e.g. Grant 1986; Boag 1988; Grant and Grant 1989). Seabird populations at the Galápagos also are affected by El Niño events, with numbers of Flightless cormorants, *Nannopterum harrisi*, and Galápagos penguins, *Spheniscus mendiculus*, being reduced to ca. 400 each as a result of the 1982–1983 El Niño (Castillo 1984). This suggests that these species have been subjected to frequent bottlenecks, and might be expected to have very little genetic diversity. It is particularly likely for the cormorant, given its very limited movements (rarely > 1 km, Harris 1979) and high levels of inbreeding (Tindle 1984).

2.4.3 Learning from Recent Extinctions

The danger of considering extant taxa with small populations is that success stories predominate; there have been no recent extinctions of bird taxa above the population level which cannot at least in part be attributed to human activities (Greenway 1967; King 1978). However, it remains useful to consider the few well-documented extinctions of natural populations. The best-known example is perhaps that of the heath hen, *Tympanuchus c. cupido* (Gross 1932). Formerly widespread and abundant in eastern North America, it was reduced by hunting pressure to only 200 birds on Martha's Vineyard Island in 1890. Further losses were halted in 1908, when a reserve was created for the last 50 birds, and captive breeding and translocations were attempted. By 1915 there were approximately 2000 health hens on Martha's Vineyard. However, on 12 May 1916, a fire burnt most of the reserve, killing many females on their nests. The 1916 winter was harsh and saw a large influx of avian predators, causing additional mortality. By the spring of 1917 there were only 150 health hens, most of which were males. The final blow was dealt in 1920 when a poultry disease, brought to the island by introduced turkeys, spread among the population. By 1927 there were only 13 birds, 11 of which were males. Despite intensive efforts, the last bird disappeared in 1931.

Numerous factors played a role in the extinction of the heath hen, in what is a classic example of the extinction vortex (Gilpin and Soulé 1986). An additional factor that may have exacerbated the situation is that the Martha's Vineyard population possibly was introduced to the island (Gross 1932). This means that it may have undergone a severe bottleneck prior to the events outlined above. Simberloff (1988) suggests that the population was doomed before 1908, apparently because it became extinct despite cessation of the systematic human pressures. This may be unnecessarily gloomy, given the recovery of species that have been in equally dire straits. For example, the Laysan finch, *Telespyza cantans*, population crashed to only a few dozen birds in 1923 after overgrazing by rabbits, but subsequently recovered to its present population of ca. 10 000 birds (Fleischer et al. in press). Moreover, the species has shown considerable morphological adaptability when introduced to novel habitats (Conant 1988).

There are virtually no other well-documented accounts of extinctions of small populations of birds. Pettersson (1985) recorded the disappearance of an isolated population of Middle Spotted Woodpeckers, *Dendrocopos medius*. After persisting for at least 30 years with a population size of 25–35 individuals, the population died out within 6 years, without significant habitat alteration. Extinction occurred despite supplementary feeding and protection from human disturbance, and seemed to result from a combination of heavy winter mortality and low reproductive success. The proportion of mated birds remained constant throughout the decline, despite very low numbers and fragmented habitat. However, there may have been some difficulty finding mates, because only 62% of birds were mated. Pettersson (1985) invoked inbreeding depression to account for the low reproductive success, although there is little evidence that egg hatchability was significantly reduced. Alternatively, demographic stochasticity following the series of cold winters could account for the extinction.

2.5 Conclusions

Despite the threat of massive loss of biological diversity and the concomitant burgeoning interest in the novel science of conservation biology (Simberloff 1988), there has been relatively little advance during the past decade in understanding the extinction process. Sophisticated models such as PVA have teased apart many of the factors affecting specific populations, but few generalized patterns have emerged. Further research on the biological attributes predisposing certain species to the threat of extinction is urgently required (Lande 1988). However, until we are better able to predict the persistence of populations and communities, rules of thumb such as MVP are useful stop-gaps for practical management.

We have attempted to review the factors affecting the persistence of natural populations of birds in relation to the size of the population, given the absence of systematic pressures such as habitat destruction or direct exploitation. Our main aim has been to contrast the empirical evidence for the persistence and performance of small populations of birds with that predicted by theory. For simplicity, we have focussed on single populations, and have not touched upon such important spatial aspects such as habitat fragmentation, metapopulation theory (e.g. Hanski and Gilpin 1991) and the associated trade-offs that determine whether it is better to maintain one large or many small populations (e.g. Lacy 1987; Shaffer 1987; Gilpin 1991).

Empirical observations of bird populations suggest that population size is the major determinant of the short-term risk of extinction. Demographic and environmental variability together are likely to cause the extinction of small population sizes (<20 breeding pairs) within 100 years. However, at larger population sizes, other factors such as differences between species in terms of demographic parameters, intrinsic behavioural traits, and interdependence with other taxa/communities preclude the detection of universal thresholds above which the risk of extinction is greatly reduced. Some species have persisted for evolutionary periods at effective population sizes of fewer than 100 birds, but they probably represent only a very small proportion of success stories among large numbers of previous extinctions. For most species, a much larger population size is probably needed to ensure a reasonable likelihood of survival in the medium- to long term. Theoretical and empirical evidence together suggest that a population of the order of 5000 individuals is needed to survive environmental stochasticity, given a single population and no active management (Soulé 1987b; Thomas 1990).

Genetic stochasticity places theoretical constraints on population viability, but its importance needs further verification. The somewhat conflicting evidence on inbreeding depression indicates that this may be less of a problem than has been stated (cf. Soulé 1980; Ralls et al. 1986; Lande 1988). Some species probably could tolerate high levels of inbreeding, but monitoring of key reproductive indicators should be undertaken for threatened populations.

The question of maintaining genetic diversity is more problematic. There are several examples of large mammal populations that persist despite having very little genetic variation (e.g. polar bear, *Thalarctos maritimus*, Allendorf et al. 1979; cheetah, *Acinonyx jubatus*, O'Brien et al. 1986), but their evolutionary potential presumably is equally limited. We cannot predict the magnitude of evolutionary responses required for lineages to survive future events, so the ideal situation would be to preserve as much genetic diversity as possible. However, the evolutionary potential of a large proportion of species already has been compromised by the destruction and fragmentation of natural habitats.

A variance effective population size of 500 may be the best minimum requirement (cf. Lande and Barrowclough 1987), and this typically means a total population in the low thousands, similar to that needed to survive environmental variability. However, smaller populations retain considerable evolutionary potential, and should not be discarded as having little or no evolutionary role. The "bottom line" for managers is that populations as small as several hundred individuals have a very good chance of survival, particularly given monitoring of the population's demographic parameters to give early warning of impending problems. Active steps could be taken to help small populations avoid severe bottlenecks, with the concomitant high risk of extinction.

Acknowledgements. Financial assistance was provided by the Werner Reimers Foundation, the South African Department of Environment Affairs, the Southern African Nature Foundation, the Foundation for Research and Development, and the University of Cape Town. Additional financial and logistical support was received from Tristan Investments (Pty) Ltd, the American Natural History Museum's Chapman Grant Fund, and the Wildlife Society of Southern Africa. Professor W.J. Bond provided useful comments on an earlier draft.

References

Allendorf FW, Christiansen FB, Dobson T, Eanes WF, Frydenberg O (1979) Electrophoretic variation in large mammals. I. The Polar Bear, *Thalarctos maritimus*. Hereditas 91: 19–22

Baker AJ, Dennison MD, Lynch A, Le Grand G (1990) Genetic divergence in peripherally isolated populations of Chaffinches in the Atlantic Islands. Evolution 44: 981–999

Barrowclough GF (1978) Sampling bias in dispersal studies based on finite area. Bird-Banding 49: 333–341

Barrowclough GF (1980) Gene flow, effective population size, and genetic variance components in birds. Evolution 34: 789–798
Bateson P (1982) Preferences for cousins in Japanese Quail. Nature 295: 236–237
Beardmore JA (1983) Extinction, survival, and genetic variation. In: Schoenewald-Cox CM, Chambers SM, Macbryde B, Thomas WL (eds) Genetics and conservation: a reference manual for managing wild animal and plant populations. Benjamin/Cummings, Menlo Park, CA, pp 125–151
Belovsky GE (1987) Extinction models and mammalian persistence. In: Soulé ME (ed) Viable populations for conservation. Cambridge University Press, Cambridge, pp. 35–57
Boag PT (1988) The genetics of islands birds. Proc Int Ornithol Congr 19: 1550–1563
Bolger DT, Alberts AC, Soulé ME (1991) Occurrence of bird species in habitat fragments: sampling, extinction, and nested species subsets. Am Nat 137: 155–166
Cade TJ (1983) Hybridization and gene exchange among birds in relation to conservation. In: Schoenewald-Cox CM, Chambers SM, Macbryde B, Thomas WL (eds) Genetics and conservation: a reference manual for managing wild animal and plant populations. Benjamin/Cummings, Menlo Park, CA, pp 288–309
Castillo CAV (1984) Breve nota sobre el impacto de El Niño 1982–83 en las poblaciones de aves marinas en las Galápagos. Bol ERFEN 11: 18–19
Charlesworth D, Charlseworth B (1987) Inbreeding depression and its evolutionary consequences. Annu Rev Ecol Syst 18: 237–268
Collar NJ, Stuart SN (1985) Threatened birds of Africa and related islands. International Council for Bird Preservation and International Union for Conservation of Nature and Natural Resources, Cambridge
Conant S (1988) Geographic variation in the Laysan finch (*Telespyza cantans*). Evol Ecol 2: 270–282
Craig JL (1991) Are small populations viable? Proc Int Ornithol Congr 20: 2546–2552
Craig JL, Jamieson IG (1988) Incestuous mating in a communal bird: a family affair. Am Nat 131: 58–70
Crow JF, Kimura M (1970) An introduction to population genetics theory. Harper & Row, New York
Diamond JM (1984) "Normal" extinctions of isolated populations. In: Nitecki MH (ed) Extinctions. University of Chicago Press, Chicago pp 191–245
Eisen EJ (1975) Population size and selection intensity effects on long-term selection response in mice. Genetics 79: 305–323
Ewens WJ, Brockwell PJ, Gani JM, Resnick SI (1987) Minimum viable population size in the presence of catastrophes. In: Soulé ME (ed) Viable populations for conservation. Cambridge University Press, Cambridge, pp. 59–68
Fleischer RC, Conant S, Morin MP (in press) Genetic variation in native and translocated populations of the Laysan finch (*Telespyza cantans*). Heredity
Franklin IR (1980) Evolutionary change in small populations. In: Soulé ME, Wilcox BA (eds) Conservation biology: an evolutionary-ecological perspective. Sinauer, Sunderland, MA, pp. 135–150
Freed LA, Conant S, Fleischer RC (1987) Evolutionary ecology and radiation of Hawaiian passerine birds. Trends Ecol Evol 2: 196–203
Gibbs HL, Grant PR (1989) Inbreeding in Darwin's medium ground finches (*Geospiza fortis*). Evolution 43: 1273–1284
Gill FB (1980) Historical aspects of hybridization between blue-winged and golden-winged warblers. Auk 97: 1–18
Gilpin ME (1987) Theory vs. practice. Trends Ecol Evol 2: 169
Gilpin ME (1991) The genetic effective size of a metapopulation. Biol J Linn Soc 42: 165–175
Gilpin ME, Soulé ME (1986) Minimum viable populations: processes of species extinction. In: Soulé ME (ed) Conservation biology: the science of scarcity and diversity. Sinauer, Sunderland, MA, pp. 19–34
Goodman D (1987a) How do any species persist? Lessons for conservation biology. Conserv Biol 1: 59–62
Goodman D (1987b) The demography of chance extinction. In: Soulé ME (ed) Viable populations for conservation. Cambridge University Press, Cambridge, pp. 11–34

Grant PR (1986) Ecology and evolution of Darwin's finches. Princeton University Press, Princeton
Grant BR, Grant PR (1989) Evolutionary dynamics of a natural population: the large cactus finch of the Galápagos. University of Chicago Press, Chicago
Grant PR, Grant BR (1992) Hybridization of bird species. Science 2566: 193–197
Greenway JC (1967) Extinct and vanishing birds of the world. Dover, New York
Greenwell GA, Emerick C, Biben M (1982) Inbreeding depression in mandarin ducks: a preliminary report on some continuing experiments. Avic Mag 88: 145–148
Greenwood PJ (1987) Inbreeding, philopatry and optimal outbreeding in birds. In: Cooke F, Buckley PA (eds) Avian genetics: a population and ecological approach. Academic Press, London, pp 207–222
Gross AO (1932) *Tympanuchus cupido cupido* (Linnaeus) heath hen. In: Bent AC (ed) Life histories of North American gallinaceous birds. US Nat Mus Bull 162: 264–280
Gyllensten U, Ryman N, Saether T (1985) Genetic divergence between willow grouse (*Lagopus lagopus*) and rock ptarmigan (*Lagopus mutus*) and the genetic structure of Scandanavian grouse populations. Hereditas 102: 47–55
Handford P (1980) Heterozygosity at enzyme loci and morphological variation. Nature 286: 261–262
Hanski I, Gilpin ME (1991) Metapopulation dynamics: brief history and conceptual domain. Biol J Linn Soc 42: 3–16
Harris MP (1979) Population dynamics of the flightless cormorant, *Nannopterum harrisi*. Ibis 121: 135–146
Harris RB, Maguire LA, Shaffer ML (1987) Sample sizes for minimum viable population estimation. Conserv Biol 1: 72–76
Höflich O (1984) Climate of the South Atlantic Ocean. In: van Loon H (ed) World survey of climatology, vol 15. Climates of the oceans. Elsevier, Amsterdam, pp 1–195
Jones HL, Diamond JM (1976) Short-time-base studies of turnover in breeding bird populations on the California Channel Islands. Condor 78: 526–549
Karr JR (1982a) Population variability and extinction in the avifauna of a tropical land bridge island. Ecology 63: 1975–1978
Karr JR (1982b) Avian extinction on Barro Colorado Island, Panama: a reassessment. Am Nat 119: 220–239
Karr JR (1990) Avian survival rates and the extinction process on Barro Colorado Island, Panama. Conserv Biol 4: 391–397
Kilpatrick CW (1981) Genetic structure of insular populations. In: Smith MH, Joule J (eds) Mammalian population genetics. University of Georgia Press, Athens, GA, pp 28–59
King WB (1978) Endangered birds of the world and current efforts towards managing them. In: Temple SA (ed) Endangered birds: management techniques for preserving threatened species. University of Wisconsin Press, Madison, pp 9–17
King WB (1981) Endangered birds of the world. Smithsonian Institution Press, Washington
Koenig WD, Pitelka FA (1979) Relatedness and inbreeding avoidance in the communally nesting acorn woodpecker. Science 206: 1103–1105
Lacy RC (1987) Loss of genetic diversity from managed populations: interacting effects of drift, mutation, immigration, selection, and population subdivision. Conserv Biol 1: 143–158
Lande R (1987) Extinction thresholds in demographic models of territorial populations. Am Nat 130: 624–635
Lande R (1988) Genetics and demography in biological conservation. Science 241: 1455–1460
Lande R, Barrowclough GF (1987) Effective population size, genetic variation, and their use in population management. In: Soulé ME (ed) Viable populations for conservation. Cambridge University Press, Cambridge, pp 87–123
Leigh EG (1981) The average lifetime of a population in a varying environment. J Theor Biol 90: 213–239
Long JL (1981) Introduced birds of the world. Universe Books, New York
Lovejoy TE, Rankin JM, Birregaard RO Jr, Brown KS Jr, Emmons LH, Van Der Voort ME (1984) Ecosystem decay of Amazon forest remnants. In: Nitecki MH (ed) Extinctions. University of Chicago Press, Chicago, pp 295–325

MacArthur RH, Wilson EO (1967) The theory of island biogeography. Princeton University Press, Princeton
Mayfield HF (1983) Kirtland's Warbler, victim of its own rarity? Auk 100: 974–976
Motro U, Thomson G (1982) On heterozygosity and the effective size of populations subject to size changes. Evolution 36: 1059–1066
Newmark WD (1991) Tropical forest fragmentation and the local extinction of understory birds in the eastern Usambara Mountains, Tanzania. Conserv Biol 5: 67–78
Newton I (1989) Synthesis. In: Newton I (ed) Lifetime reproductive success in birds. Academic Press, London, pp 441–469
O'Brien SJ, Wildt DE, Bush M (1986) The cheetah in genetic peril. Sci Am 254: 84–92
Payne RB (1990) Natal dispersal, area effects, and effective population size. J Field Ornithol 61: 396–403
Pettersson B (1985) Extinction of an isolated population of the middle spotted woodpecker, *Dendrocopos medius* (L.) in Sweden and its relation to general theories on extinction. Biol Conserv 32: 335–353
Pimm SL, Redfearn A (1988) The variability of natural population densities. Nature 334: 613–614
Pimm SL, Jones HL, Diamond JM (1988) On the risk of extinction. Am Nat 132: 757–785
Preece RC, Bennett KD, Carter JR (1986) The Quaternary palaeobotany of Inaccessible Island (Tristan da Cunha group). J Biogeogr 13: 1–33
Ralls K, Harvey PH, Lyles AM (1986) Inbreeding in natural populations of birds and mammals. In: Soulé ME (ed) Conservation biology: the science of scarcity and diversity. Sinauer, Sunderland, MA, pp 35–56
Rowley I, Russell E, Brooker M (1986) Inbreeding: benefits outweigh costs. Anim Behav 34: 939–941
Schoener TW (1976) The species-area relation within archpelagos: models and evidence from island land birds. Proc Int Ornithol Congr 16: 629–642
Schonewald-Cox CM, Chambers SM, Macbryde B, Thomas WL (eds) (1983) Genetics and conservation: a reference manual for managing wild animal and plant populations. Benjamin/Cummings, Menlo Park, CA
Shaffer ML (1981) Minimum population sizes for species conservation. BioScience 31: 131–134
Shaffer ML (1987) Minimum viable populations: coping with uncertainty. In: Soulé (ed) Viable populations for conservation. Cambridge University Press, Cambridge, pp 69–86
Shaffer ML (1990) Population viability analysis. Conserv Biol 4: 39–40
Shields WM (1982) Philopatry, inbreeding, and the evolution of sex. State University of New York Press, Albany
Simberloff D (1988) The contribution of population and community biology to conservation science. Annu Rev Ecol Syst 19: 473–511
Sittmann K, Abplanalp H, Fraser RA (1966) Inbreeding depression in Japanese quail. Genetics 54: 371–379
Soulé ME (1980) Thresholds for survival: maintaining fitness and evolutionary potential. In: Soulé ME, Wilcox BA (eds) Conservation biology: an evolutionary-ecological perspective. Sinauer, Sunderland, MA, pp. 151–169
Soulé ME (1983) What do we really know about extinction? In: Schoenewald-Cox CM, Chambers SM, Macbryde B, Thomas WL (eds) Genetics and conservation: a reference manual for managing wild animal and plant populations. Benjamin/Cummings, Menlo Park, CA, pp 111–124
Soulé ME (ed) (1986) Conservation biology: the science of scarcity and diversity. Sinauer, Sunderland, MA
Soulé ME (ed) (1987a) Viable populations for conservation. Cambridge University Press, Cambridge
Soulé ME (1987b) Where do we go from here? In: Soulé ME (ed) Viable populations for conservation. Cambridge University Press, Cambridge, pp 175–183
Soulé ME, Bolger DT, Alberts AC, Wright J, Sorice M, Hill S (1988) Reconstructed dynamics of rapid extinctions of chaparral-requiring birds in urban habitat islands. Conserv Biol 2: 75–92

Templeton AR, Read B (1983) The elimination of inbreeding in a captive herd of Speke's Gazelle. In: Schoenewald-Cox CM, Chambers SM, Macbryde B, Thomas WL (eds) Genetics and conservation: a reference manual for managing wild animal and plant populations. Benjamin/Cummings, Menlo Park, CA, pp 241–261

Terborgh J, Winter B (1980) Some causes of extinction. In: Soulé ME, Wilcox BA (eds) Conservation biology: an evolutionary-ecological perspective. Sinauer, Sunderland, pp 119–133

Thomas CD (1990) What do real population dynamics tell us about minimum viable population sizes? Conserv Biol 4: 324–327

Tindle R (1984) The evolution of breeding strategies in the flightless cormorant (*Nannopterum harrisi*) of the Galápagos. Biol J Linn Soc 21: 157–164

Tomlinson C, Mace GM, Black JM, Hewston N (1991) Improving the management of a highly inbred species: the case of the white-winged wood duck, *Cairina scutulata*, in captivity. Wildfowl 42: 123–133

Tracy CR, George TL (1992) On the determinants of extinction. Am Nat 139: 102–122

Van Noordwijk AJ, Scharloo W (1981) Inbreeding in an island population of the Great Tit. Evolution 35: 674–688

Vasquez CG, Bohren BB (1982) Population size as a factor in response to selection for eight-week body weight in white leghorns. Poult Sci 61: 1273–1278

Walter HS (1990) Small viable population: the red-tailed hawk of Socorro Island. Conserv Biol 4: 441–443

Weller MW (1980) The island waterfowl. Iowa State University Press, Ames

Wilcox BA (1980) Insular ecology and conservation. In: Soulé ME, Wilcox BA (eds) Conservation biology: an evolutionary-ecological perspective Sinauer, Sunderland, pp 95–117

Willis EO (1974) Populations and local extinctions of birds on Barro Colorado Island, Panama, Ecol Monogr 44: 153–169

Yang SY, Patton JL (1981) Genic variability and differentiation in the Galápagos finches. Auk 98: 230–242

Zimmerman BL, Bierregaard RO (1986) Relevance of the equilibrium theory of island biogeography and species-area relations to conservation with a case from Amazonia. J Biogeogr 13: 133–143

3 Small World Populations in Birds: An Attempt of a Brief General Survey

E. BEZZEL

3.1 Introduction

The class Aves is said to be the best-known or at least one of the best-known animal taxa. However, estimates of wold population size cover only less than 25% of the 1029 threatened bird species listed by Collar and Andrew (1988). This means that over 11% of the approximately 9000 living bird species are to varying degrees at risk from global extinction. The number of counts and estimates of local or regional populations, however, is almost innumerable, but most of them refer to only a very small fraction of the species' world population. Therefore local population size and development may be influenced by immigration, depending on the strength of isolating mechanisms. Those systems fitting more or less with the concept of metapopulation (e.g. Hanski and Gilpin 1991) with "rescue-effects" (Brown and Kodric-Brown 1977) not rarely work over considerable distance, as birds are highly mobile organisms.

The basis for comparing size and change of small world populations in birds as units of "minimum populations" in a brief overall survey is therefore not only rather small, but also presumably a heavily biased sample. Some biota, e.g. primary forests in different climatic zones, are clearly under-represented, as there are great methodical problems in collecting reliable data on actual population size (e.g. Diamond and Lovejoy 1985; Fanshawe and Bennun 1991; Green and Hirons 1991). Therefore, conclusions must be drawn very cautiously even if we only try to give a rough descriptive comparison of the situation for a number of populations of various bird taxa representing different ecological adaptions or life strategies.

The data presented here refer to actual population sizes. Whether they can be interpreted as Minimum Viable Populations (MVP, e.g. Shaffer 1981; Ryan & Siegfried, Chap. 2, this vol.) needs further studies. Some of the very small populations seem to be rather stable (see also Ryan and Siegfried, Chap. 2, this vol.), in some others protection of habitat or special management techniques resulted in a considerable increase in individuals or was able to prevent a further decline.

3.2 Material and Methods

3.2.1 Problems in Data Selecting and Weighing Up

This chapter includes 254 world populations of birds for which data were collected from various sources, and which are at least taxonomically treated as subspecies or semispecies, thus indicating a degree of isolation to develop unique phenotypic characters. The upper limit considered is 50 000 individuals, regardless of their dispersion (colonial, inhabiting one small area, or patchily dispersed over a wide range). Thus also metapopulations or more or less isolated populations belonging to one taxon have been considered as one unit. Only in the case of Waldrapp ibis (*Geronticuss eremita*) were the two populations of Morocco and the Near East treated separately.

The number of populations covered here exceeds the 202 species up to a size of > 100 000 individuals collected by Green and Hirons (1991) because I have included not only threatened species but also subspecies and species with small populations whether "threatened" or not.

The size of a particular population considered here does not always refer to the most recent estimate or census. If there are several data over a longer period, mostly the minimum figure has been taken and/or a figure from earlier years when there was at least one more recent census available allowing measurement or estimation of a population trend after the time for which the size was measured. As many methodical problems of reliable estimates of bird populations have to be taken into account (e.g. Green & Hirons 1991), the population size is given in size classes, the ranges of which – as in many other classifications in this chapter – are chosen rather arbitrarily.

Therefore, for example, it is not possible to give a frequency distribution of the size classes for small or "minimum" populations. The nearly normal frequency distribution of estimated population size culminating in the size class 101–1000 individuals in Green and Hirons (1991) may be highly biased, as estimates for larger populations (e.g. > 50 000 individuals) are presumably underrepresented. Sometimes early and more or less preliminary figures published for some threatened taxa were too low according to later more exact counts and research. Only for a small number of populations are long-run census data available, so that trend estimates are mostly based on a few data or reliable assumptions. Here, only trend estimates running over at least 10 years are considered.

3.2.2 Grouping the Taxa of Birds

The data on bird populations collected were divided into four different groups which only very roughly correspond with taxonomic classification. "Life styles" seem to be more appropriate for understanding the situation in small populations, as they can explain demographic and environmental effects. As, however, the total of 254 populations is too low for separating many well-defined groups,

Table 3.1. Population size and distribution of small seabird populations (n)

Population (ind.)	Distribution[a] n	1	2	3	Size of breeding range (km^2) –50	51–1000	1001–10000	>10000
1–50	3	3			2	1		
51–500	6	4	2		2	2	1	1
501–1000	5	4	1		1	2	2	
1001–5000	16	11	4	1	5	9		2
5001–10000	5	2	2			3		1
10001–50000	12	4	4	4	3	3	3	3
1–1000	14	79%	21%	–	36%	36%	21%	7%
1001–50000	33	52%	33%	15%	24%	48%	9%	18%
Total	47	60%	30%	30%	28%	45%	13%	15%

[a]Distribution: 1, one island or island group; 2, at least 2 islands or island groups separated by >100 km or islands and coasts; 3, continent. Percentages rounded.

the criteria must be selected within a wide range of variability. Thus, the different taxa lumped together in one group mostly exhibit only approximately some of the main criteria of seabirds, medium-sized and large landbirds, small land birds and birds of wetlands (mostly freshwater and/or sodaic lakes).

1. Seabirds (Table 3.1) depend on land to breed; their breeding grounds are generally restricted to islands and/or shores. Most of them breed colonially or semicolonially, and the breeding sites are often relatively small, whereas the feeding grounds, whether inshore or offshore, as well as the ranges outside the breeding season, mostly comprise a relatively large area, in some species (e.g. albatrosses, shearwaters) even great parts of the ocean. The life history is characterized by a very high individual longevity in combination with late sexual maturity and in some species with a high amount of non-breeding individuals per season. Therefore, fitness and lifetime reproduction are mainly related to longevity, as the annual offspring production is very low. The length of generation defined as the time required for a population to replace itself is generally great. There is, however, a wide range of variability between taxa and size classes (e.g. terns, gulls vs. albatrosses, shearwaters or auks).

2. Larger landbirds (Table 3.2) form a very heterogeneous group as they include carnivores, insectivores, and predominantly herbivores, as well as different size classes from stork or crane down to dove. Most of the taxa involved show a relatively long individual longevity or low annual adult mortality combined with very low (e.g. large raptors) up to rather high (e.g. grouse) annual production. In the latter case, the losses of eggs and chicks, as well as juvenile mortality, are high. Predominantly in larger birds of this group (but also differing between larger taxonomic levels such as orders and/or families), the age at first breeding is >1 years, the generation time mostly shorter than in seabirds but frequently much longer than in small landbirds. Especially birds of the open land (e.g. cranes, bustards) and large raptors need large territories or home ranges for individuals, breeding units or groups all the year round. Therefore

Table 3.2. Population size and distribution of small populations in large and medium-sized landbirds (see Table 3.1)

Population (ind.)	Distribution n	1	2	3	Size of breeding range (km^2) −50	51–1000	1001–10 000	>10 000
1–50	9	7		2	3	5		1
51–500	24	16	1	7	5	12		7
501–1000	13	4	1	8	2	3	1	7
1001–5000	20	7	2	11	1	11	1	7
5001–10 000	9	1		8	1			8
10 001–50 000	6			6				6
1–1000	46	59%	4%	37%	22%	43%	2%	33%
1001–50 000	35	23%	6%	71%	6%	31%	3%	60%
Total	81	43%	5%	52%	14%	39%	3%	45%

density is mostly low; even for small populations large areas of suitable habitat are a prerequisite. Some species are long-distance migrants.

3. The bulk of small landbirds (Table 3.3) form species and subspecies of the order Passeriformes. Many of them depend on forests or at least on higher vegetation. Normally, the individual lifetime is short, annual adult and juvenile mortality high. The fitness is mainly related to the number of annual offspring, the generation time often being very short. As in many species, breeding units need only small territories, while the overall density at least in optimal habitats can be very high, particularly in birds living in higher vegetation or in different vegetation layers. In contrast to higher latitudes and some island populations, the density of species living in tropical forests often seems to be surprisingly low. Forest species, however, are mostly excluded in this chapter as data on population size are lacking. Within larger areas the distribution sometimes is rather patchy. Very few long-distance migrants are included (e.g. Kirtland's warbler, *Dendroica kirtlandii*).

Table 3.3. Population size and distribution of small population in small landbirds (see Table 3.1)

Population (ind.)	Distribution n	1	2	3	Size of breeding range (km^2) −50	51–1000	1001–10 000	>10 000
1–50	19	18		1	14	4		1
51–500	40	34		6	16	22	1	1
501–1000	9	7		2	4	3		2
1001–5000	19	13		6	7	9	1	2
5001–10 000	4	3		1	2	2		
10 001–50 000	3	2		1	1	2		
1–1000	68	87%		13%	50%	43%	1%	6%
1001–50 000	26	69%		31%	31%	38%	4%	8%
Total	94	82%		18%	47%	45%	2%	6%

Table 3.4. Population size and distribution in birds of inland wetlands (see Table 3.1)

Population (ind.)	Distribution n	1	2	3	Size of breeding range (km^2) −50	51–1000	1001–10000	>10000
1–50	3	2		1	1	1		1
51–500	8	2		6	1	6		1
501–1000	2	1		1		1		1
1001–5000	10	2	2	6	1			9
5001–10000	2	1		1				2
10001–50000	7		1	6			1	6
1–1000	13	38%		62%	15%	62%		23%
1001–50000	19	16%	16%	68%	5%		5%	89%
Total	32	25%	9%	66%	9%	25%	3%	63%

4. The fourth group, birds of wetlands (Table 3.4), is rather heterogeneous again. Birds of wetlands generally share a discontinuously distributed habitat or a single isolated breeding site limiting the population size. Here, populations living all the year round on or at inland waterbodies or coastal lagoons are summed up, e.g. grebes, ducks and some waders (waders breeding in steppes or dry to moist meadows like curlews, *Numenius*, are considered as landbirds). Most of the populations have a rather long generation time; the annual production is medium to high.

3.3 Results and Conclusions

3.3.1 Distribution

The frequency of small world populations compiled in Tables 3.1–3.4 is unevenly distributed over the classical zoogeographic regions (e.g. Mountfort 1988). Their longitudinal and latitudinal distribution roughly matches the Endemic Bird Areas (EBA) described in detail by Bibby et al. (1992). Besides small and larger oceanic islands from low northern to high southern latitudes, continental populations are concentrated in areas from southern North to southern South America, in Subsaharan Africa, from continental Southeast Asia to New Guinea and Australia; only a few populations are confined to the temperate and northern Holarctic region.

3.3.2 Population Size and Dispersion

In seabirds (Table 3.1) populations up to 1000 individuals breed on islands, most of them at only one site. However, the number of cases documented is rather low. For most populations, even those existing in greater numbers (>50000 individuals), breeding sites cover a very small land area giving place for dense colonies only; some species are well adapted to steep cliffs and breed only in particular "layers" of a vertical cliff. Thus populations of up to 50 000 individuals included in Table 3.1 are mainly concentrated on islands, 28% even on rather small ones. About 15% are found only on continental shores. The greater size classes of land areas shown in Table 3.1 are by no means a very meaningful description of the breeding range, as the birds search for food at sea. Moreover, only very few and small sites normally provide suitable breeding conditions such as good access to the nests (especially for flightness penguins or for long-winged albatrosses depending on prevailing winds), ledges and/or holes for the eggs and more or less altricial chicks, and high productive feeding grounds in adjacent waters, which can be reached daily. Seabird colonies on a particular site therefore often have a long tradition, even in species which travel over great distances over the oceans during their premature life and/or outside breeding season.

Small populations of medium-sized and large landbirds are clearly more continental than seabirds, but a remarkably high percentage is restricted to a single island or island group (Table 3.2). The very low number of populations inhabiting different separated islands demonstrates the isolating effect of the sea, inhibiting a high gene flow over greater distances. Contrary to the case of seabirds, the breeding ranges of landbirds include feeding areas as well (even in colonial species) and the percentage of large ranges particularly for populations >1000 individuals is much higher. The 45% of the populations listed in Table 3.2 dispersed over areas $>10\,000\,\mathrm{km}^2$, however, indicate a very low overall density which means that either individuals/pairs need a large home range or the whole population is discontinuously distributed, mostly due to habitat fragmentation.

The rather high number of populations below 500 individuals listed for small landbirds (Table 3.3) may be an artefact because for greater populations census data are lacking, particularly in forest species. The skewed frequency distribution of data sets over different habitats may also be responsible for the high amount of island taxa in comparison to continents, where presumably much more small world populations may be found in valleys of high mountains or restricted forest patches, nowadays very often as a consequence of fragmentation by human activities. All the small insular world populations of smaller landbirds, however, are restricted to one oceanic island or island group; in adjacent islands, closely related taxa are often phenotypically different so that they may be treated as subspecies at least. Generally, greater distances to continents or between islands seem to have a strong isolating effect. Some small populations in islands, however, are the result of adaptive radiation for which the Darwin's finches, Geospizinae, of Galapagos and the Hawaiian Honeycreepers, Derpanidinae, provide well-known examples. In contrast to larger landbirds, the populations listed in Table 3.3 often inhabit a very small area (see below).

This can also be stated for birds breeding in/on lagoons or inland waters (Table 3.4) if they are restricted to single systems. These populations in particular comprise the highest amount of continental taxa in the four groups separated here, among which, as in insular larger landbirds (Table 3.2), some flightless or poorly flying species occur.

In total, 34 (13%) of the small world populations summed up here consist of not more than 50 and 78 (31%) between 50 and 500 individuals, nearly 53% of them being small landbirds (mostly Passerines). However, a possible bias should be kept in mind in relating to these figures.

The rarest birds which may have survived free living to the present comprise not more than 20 individuals, for some of which only single individuals could be traced in recent years. Among those extremely rare taxa could be counted: on Hawaii Kauai Oo *Moho braccatus*, Olomao *Myadestes lanaiensis*, Puaiohi *Myadestes palmeri*; on the Marquesas a subspecies of the Marquesas monarch, *Pomara mendozae nukhivae*; on Mauritius pink pigeon, *Nesoenas myeri*, Mauritius parakeet, *Psittacula eques*; in the northernmost Nearctic region Eskimo curlew, *Numenius borealis*; in the eastern Palaearctic region crested ibis *Nipponia nippon*. Among those populations which have declined at least for a short period to a size between 20 and 50 individuals, 19 are restricted to rather small oceanic islands, 2 to New Zealand and 1 to continent Oriental region.

3.3.3 Population Trends

Small world populations may often be the result of heavy decline, either long-term or caused within a relatively short time by human influences (habitat destruction or introduced foreign species or domestic animals on islands, deliberately or by chance). Before a population of 1000 individuals or well below was reached, in 111 (90%) out of 124 cases a decline has at least been estimated, among populations of >1000 to 50000 individuals ($N = 68$) a decrease can be assumed for 43 (63%, Table 3.5). These percentages, however, may be biased, as only for few populations is the development over a longer period adequately known. If we consider very small (up to 1000 individuals) vs. small (1001–50000

Table 3.5. Tendencies of population changes is small bird populations before the size considered was reached[a]

Population (ind.)	A			B			C			D		
	+	±	−	+	±	−	+	±	−	+	±	−
1–50			2		1	7	2	4	15			3
51–500			5		2	18			21			8
501–1000		1	3	1		7	2	2	21			2
1001–5000		2	6	8	3			1	4			7
5001–10000		1	2			4	1	1	9			1
10001–50000		5	1		2	3			1		1	5
1–1000		1	10	1	3	32	4	4	56			13
1001–50000		8	9	8	5	7	1	2	14		1	13

[a] A, Seabirds; B, large landbirds; C, small landbirds; D, birds of wetlands. + Increase, ± "stable", − decrease, in ( ) to extinction.

Table 3.6. Tendencies of population change in small bird populations *after* the size considered was reached (see Table 3.5)

Population	A			B			C			D		
(ind.)	+	±	−	+	±	−	+	±	−	+	±	−
1–50			1	1	3	2(1)		6	3(1)		1	
51–500	1	1	2	2	2	10		5	7			4(2)
501–1000		3	1		2	1(1)		5	7			
1001–5000		2	6		5			1	1		3	
5001–10 000		1	2		2			1				
10 001–50 000		5	1		2	2		1			2	
1–1000	1	4	4	3	6	13(2)		13	17(1)		1	4(2)
1001–50 000		8	9		9	2		3	1		5	

individuals) populations in Table 3.5, we may find that a decline has preceded the recorded status less often in the latter (seabirds: $N = 28$, $Chi^2 = 4.4$, $p < 0.04$; larger landbirds: $N = 56$, $Chi^2 = 17.7$, $p < 0.001$); in smaller landbirds ($N = 81$), however, no difference could be detected.

The development after the size considered was reached (Table 3.6) suggests the well-known fact that smaller populations generally are more vulnerable than larger ones because of stochastic effects (and human interference). Among populations of up to 1000 individuals ($N = 66$) ca. 50% decreased (7.5% to extinction), among larger populations >1000 to 50 000 individuals ($N = 37$) only 32% (none to extinction). Due to the low figures, the difference between these two size classes is statistically not significant with exception of the larger landbirds ($N = 33$, $Chi^2 = 4.95$, $p > 0.03$). In all groups, small and even very small populations can be stable over long periods during a bottleneck situation in the absence of human interference or adverse stochastic effects (cf. Ryan and Siegfried, Chap 2, this vol.).

However, to understand the fate of small populations in detail, single cases must be analyzed separately. In particular, this is necessary to evaluate efficiently working management techniques in conservation programmes for increasing fitness. Generally, even very small populations can be preserved over rather long periods if the reasons for decline or immediate threat are known. Success has been achieved so far particularly in taxa limited to small ranges as oceanic and/or habitat islands. This may be shown in three case histories of very small populations.

The Seychelles warbler, *Acrocephalus sechellensis*, which was reduced to a population of 26 individuals in 1959, is entirely confined to Cousin Island (29 ha) of the Seychelles. Since management has been directed towards increasing suitable habitat (e.g. allowing the original vegetation to regenerate), numbers have risen to nearly 320 birds in 1982. Since 1982, the population has fluctuated around that level, suggesting that this may be the maximum carrying capacity of the island. Thus young birds were unable to establish new territories and the establishing of a second population on another island was recommended.

In 1988, 29 adult birds were transferred to Aride Island, where the birds showed nesting activity immediately following release. Four weeks after the transfer, the first young hatched (Kondeur et al. 1991).

One of the most thoroughly studied species in the Nearctic is the Kirtland's warbler *Dendroica kirtlandii*. Its distribution is extremely restricted; in historical times it inhabited an area of 136 by 160 km in northern Michigan. In 1970, its range had diminished from 4860 ha to about 2000 ha, as the habitat required must consist only of jack pine, *Pinus banksiana*, between 8 and 22 years old on sandy soil, with low undergrowth and scattered clearings. The accurate census in 1951 gave a total population of 500 pairs. Ten years later, the population was nearly the same size, but the production per pair had declined. Another decade later, the population had dropped to 210 pairs, following an intrusion of brown-headed cowbirds, *Molothrus ater*, a brood parasite which had parasited 67% of the warblers' nests. Since 1972, about 3000 cowbirds per year were removed, reducing the parasitism to below 5%. The average number of fledglings per nest tripled, but the population did not increase. It has fluctuated between 167 and 219 between 1972 and 1985, and dropped well below 200 pairs in 1987. Nevertheless, the population may be considered to have been stable for ca. 15 years. The relatively low effect of the high production achieved, which is above the average of any other American warbler, may be caused by the fact that Kirtland's warbler is a long-distant migrant wintering in the Bahamas. One thousand or 1200 birds should be migrating south each year, only about 400 surviving to return the following spring. Presumably this not only reflects an extremely high mortality; the problem of reaching a very small target in winter and in spring may also be involved (e.g. Kelly and Decapita 1982; Mountfort 1988; Green & Hirons 1991).

The Cahow, *Pterodroma cahow*, of Bermuda, found in thousands in the 16th century, had been almost exterminated by the first half of the 17th century. In 1951 a few were rediscovered, and 18 to 20 pairs of this seabird bred on a small island off Castle Harbour between 1961 and 1966. Because of contamination by DDT, only six young were reared in 1966. The population rose to a maximum of 35 pairs in 1985, which fledged 21 young. The use of DDT in the USA was banned in 1972. Moreover, the low breeding success was attributed to competition for nest sites with the aggressive White-tailed tropicbird, *Phaeton lepturus*. Artificial nest burrows were provided which exclude the competitor. The rather low increase rate may be a consequence of the long generation time typical for most seabirds.

References

Atkinson P, Peet N, Alexander J (1991) The status and conservation of the endemic bird species of Sao Tomé and Principe, West Africa. Bird Cons Int 1: 255–282

Bezzel E, Prinzinger R (1990) Ornithologie. Ulmer, Stuttgart

Bibby CJ et al. (1992) Putting biodiversity on the map: priority areas for global conservation. ICBP, Cambridge

Brown JH, Kodric-Brown A (1977) Turnover rates in insular biogeography: effect of immigration on extinction. Ecology 58: 445–449

Collar NJ, Andrew P (1988) Birds to watch. The ICBP World Check-List of threatened birds. ICBP Tech Publ 8, Cambridge

Diamond AW, Lovejoy TE (1985) Conservation of tropical forest birds. ICBP Tech Publ 4, Cambridge

Fanshawe JH, Bennun LA (1991) Bird conservation in Kenya: creating a national strategy. Bird Cons Int 1: 293–315

Green RE, Hirons GJM (1991) The relevance of population studies to the conservation of threatened birds. In: Perrins CM, Lebreton J-D, Hirons GJM (eds) Bird population studies. Oxford University Press, Oxford, pp 594–633

Hancock JA, Kushlan JA, Kahl MP (1992) Storks, ibises and spoonbills of the world. Academic Press, London

Hanski I, Gilpin ME (1991) Metapopulation dynamics: brief history and conceptual domain. Biol J Linn Soc 42: 3–16

Herremans M, Louette M, Stevens J (1991) Conservation status and vocal and morphological description of the Grand Comoro Scops owl *Otus panliani* Benson 1960. Biol Cons Int 1: 123–133

Johnson TH (1988) Biodiversity and conservation in the Carribean: profiles of selected islands. ICBP Monogr 1, Cambridge

Kelly ST, De Capita ME (1982) Cowbird control and its effect on Kirtland's warbler reproductive success. Wilson Bull 94: 363–365

Kondeur J, Bullock ID, Rands MRW (1991) Conserving the Seychelles warbler, *Acrocephalus sechellensis*, by translocation: a transfer from Cousin Island to Aride Island. Bird Cons Int 1: 177–185

Marchant S, Higgins PJ (1990) Handbook of Australian, New Zealand and Antarctic birds, vol 1. Oxford University Press, Melbourne

Mountfort G (1988) Rare birds of the world. Collins, London

Mühlenberg M, Hovestadt T, Röser J (1991) Are there minimal areas for animal populations? In: Seitz A, Loeschke V (eds), Species conservation: population-biological approach. Birkhäuser, Basel, pp 227–264

Shaffer ML (1981) Minimum population sizes for species conservation. BioScience 31: 131–134

Sibley Ch, Monroe BL (1990) Distribution and taxonomy of birds of the world. Yale University Press, New Haven

Walter HS (1990) Small viable population: the Red-Tailed Hawk of Socorro Island. Conserv Biol 4: 441–443

Watson J (1991) The Seychelles paradise flycatcher, *Terpsiphone corvina*, on La Digue: population size, habitat requirements and management. Bird Cons Int 1: 107–122

Wiley JW (1991) Status and conservation of parrots and parakeets in the Greater Antilles, Bahama Islands, and Cayman Islands. Bird Cons Int 1: 187–214

4 Genetic, Demographic, Spatial, Environmental and Catastrophic Effects on the Survival Probability of Small Populations of Mammals

H. KORN

4.1 Introduction

Because of rapid and increased insularization of remaining patches of suitable living space, the study of the ecology of small populations is of vital importance for the conservation of many species. Especially for large mammals extensive areas and the sufficient allocation of funds are required to maintain a vital breeding stock in a natural situation. Since both land and money for conservation purposes are extremely limited, priorities have to be defined. To secure as many populations and species as possible the question arises how many individuals of a population are necessary for their survival in a particular area? Too small a population is subject to sudden extirpation by natural disasters, or slower extinction due to the accumulation of deleterious alleles through inbreeding (Ralls and Ballou 1983), while very large populations of long-lived vertebrates may be prohibitively costly to maintain (Reed et al. 1986), especially in intensively managed situations.

In recent years the concept of "*M*inimum *V*iable *P*opulations" (MVP) has been elaborated and was discussed in great detail in the book of Soule (1987a). The concept aims to incorporate such fields as population dynamics, environmental variation (randomness, stochasticity), genetics, catastrophes, metapopulation structure and fragmentation (Soule 1987b). So far, the fields has been dominated by researchers who have focused on the elaboration of mathematical models. This is useful and often the only way to deal with the matter since the processes concerned may span hundreds to thousands of years and are affected by many variables. Often we are dealing with extremely low population numbers, where every individual is important for the survival of the species, like the Sumatran and Javan rhinos. They are therefore beyond the possibilities of experimental manipulations. Without going much into the mathematical details I will try to summarize the implications that these lines of thought have for the conservation of mammalian populations.

4.2 Genetic Considerations

With limited experimental evidence from the breeding of domestic animals and some theoretical considerations Franklin (1980) suggested that, in the absence of regular introductions, the effective population size (Ne) for large mammals should not be less than 50. Theoretically, Ne is the size of an "ideal" population, that is, a population of N breeding individuals, randomly mating, which produce N new individuals each generation by random union of N male and N female gemetes regarded as random samples from the population of the parent generation (Crow and Kimura 1970). The minimum number of 50 is based on the assumption that an inbreeding coefficient as high as 1% can be tolerated (Franklin 1980). This is a somewhat conservative value since animal breeders have a rule of thumb that the per generation rate of inbreeding, which is calculated by the formula 1/2Ne, should not be higher than 2 to 3%. Nevertheless, domestic animals are already adapted to close inbreeding. Higher rates of inbreeding fix deleterious genes too rapidly for selection to eliminate them from the line (Soule 1980). Also, as a rule of thumb based on domestic animals, an Ne of less than about 50 disposes a population to inbreeding depression (Franklin 1980; Soule 1980). It is important to emphasize that inbreeding accumulates at the rate which is directly related to the generation interval.

Franklin (1980) gives the example that a population of 50 randomly mating elephants will take several hundred years to reach a 10% level of inbreeding; mice will attain this figure in less than 10 years. After 20 to 30 generations the populations will have lost about 25% of their genetic variation (Soule 1980), after 100 generations the loss accumulates to more than 60% (Fig. 4.1). Random mating, however, is a rare occurrence in natural populations (Frankel 1983). Where inbreeding levels are higher, a larger population will be required to maintain an effective population size of 50, thus avoiding rapid loss of genetic variability due to inbreeding.

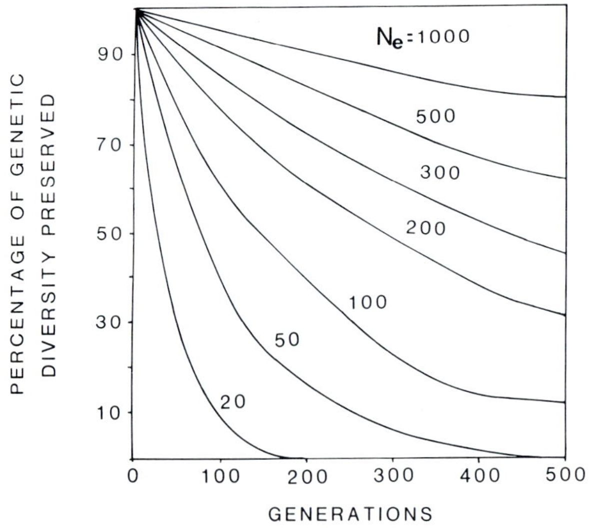

Fig. 4.1. Decline in genetic diversity resulting from random drift for various effective population sizes. (After Futuyma 1986)

Many data from natural animal populations suggest that populations with higher levels of heterozygoty are superior (Soule 1980; Allendorf and Leary 1986; Neuhäuser 1991 and references therein). Inbreeding has deleterious effects on survival and reproduction, and lowers such characters as growth rate and adult size (De Bois et al. 1990). The grizzly bear population in the Yellowstone National Park ecosystem has had a total population size of about 100 for the last five generations and an Ne of less than one-third of this during this period. This species has shown reduced litter sizes in the last generation, possibly the result of inbreeding (Chris Serveheen, pers. comm., cited in Gilpin 1987). Shaffer and Samson (1985) calculated that the Yellowstone grizzly bear population has only about a 36% chance of surviving for even 100 years and will certainly be extinct within two centuries. Anoher example: within the last 10 years the small group of wolves on Isle Royal has declined from 50 to as few as 12 individuals. One cause of the decline is considered to be reduced genetic variability which is extremely low between the animals. All animals tested are supposed to be siblings and all the remaining Isle Royal wolves are derived from a single female founder (Wayne et al. 1991). Another manifestation of loss of genetic variation is increased sensitivity to epidemic disease, because the population is not likely to have a heterogeneous immunological structure (Gilpin 1987). Close inbreeding should therefore be avoided (Senner 1980; Soule 1980). Nevertheless, it has to be mentioned that mild but regular levels of inbreeding are sufficient to eliminate inbreeding depressions (Templeton 1987). Small populations can therefore become adapted to tolerate close inbreeding, which has been shown in Speke's gazelle (*Gazella spekei*) (Ralls and Ballou 1983) and the European bison (*Bison bonatus*) in captivity (Slatis 1960). These examples provide empirical confirmation of the theoretical prediction that populations that regularly outcross should display inbreeding depressions, while normally inbred populations should display no such depressions (Templeton 1987). Although inbred populations should be free of inbreeding depression, they are more likely than outbred populations to evolve coadapted gene complexes that can yield an outbreeding depression (Shields 1982). With an inbreeding coefficient that is sooner or later higher than mutation rates, depending on the effective population size (Ne), the population will become virtually homozygous (Fig. 4.1), and will have lost much of its capacity to adapt to changing conditions (Soule 1980). This is why the basic rule to maintain a population of at least 50 breeders should only be applied for a few generations. Very small populations ($Ne < 50$) are also prone to any individual demographic stochasticity, as a skewed adult sex ratio, which was observed in the last free living population of black-footed ferrets (Brussard and Gilpin 1989). A more important consideration, however, is that populations of such small size cannot withstand the effects of environmental uncertainty. This was clearly indicated by the collapse of the above-mentioned black-footed ferret population after a distemper epizootic (Brussard and Gilpin 1989).

Most real populations do not fit the description of an effective population. Therefore, a number of formulae have been developed to estimate the ideal population size that would produce the same genetic correlations (except genotype frequencies) as those in the population under study (Chepko-Sade et al. 1987). Guidelines for the collection of complete data sets to calculate effective population sizes are given by Franklin (1980) and Chepko-Sade et al. (1987).

If not all members of the population participate in reproduction: then Ne = N × p, when N is the census number and p the fraction of N that is actually breeding. For example, if a population has 200 adults, but half of them cannot breed for whatever reason, then the Ne for that population would be 200 × 1/2 = 100.

If there is variance in progeny number: let the variance in progeny number be S2 and the census number N, then the approximation of Ne is

$$\mathrm{Ne} = \frac{4\mathrm{N}}{2 + \mathrm{S2}}.$$

In a variety of species the effective size ranged from 0.6 to 0.85 of the census numbers (Crow and Morton 1955). Effective population size normally does not reach census numbers, but under special breeding conditions, when all individuals are mated with distant relatives and variance in progeny number is eliminated, it can even be higher (Crow and Kimura 1970).

If there are unequal numbers of the two sexes: i.e., whenever the numbers of each sex contributing to the next generation are not equal, then

$$\mathrm{Ne} = \frac{1}{\frac{1}{4\mathrm{Nem}} + \frac{1}{4\mathrm{Nef}}},$$

where Nem and Nef are the number of effective males and females, respectively. This is particularly important in species which maintain stable harems like deer, seals, vicuñas, zebras, etc. (Fig. 4.2). For example, a breeding population in which

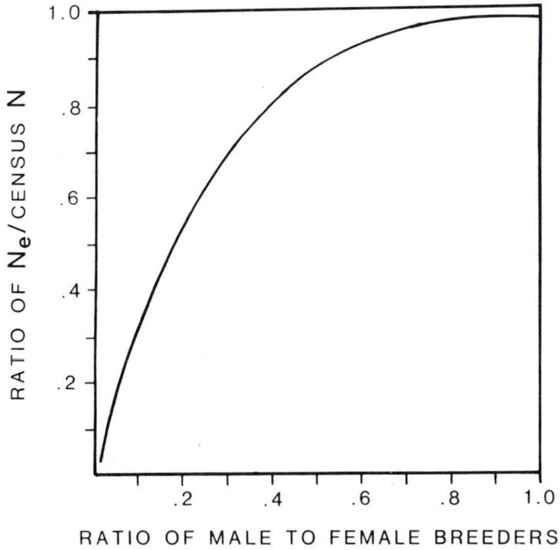

Fig. 4.2. The relationship between effective population size and census population size as a function of the sex ratio of actual breeders. As polygamy increases, effective population size decreases. (After Chepko-Sade et al. 1987)

there are 90 effective females but only 10 males has a total effective size of 36, not 100. A population of 10 males and 1000 females loses genetic variation at the same asymptotic rate as a population of 20 males and 20 females (Ewens et al. 1987).

If there is fluctuation in population size: i.e., if population size varies from generation to generation, the effective number is the harmonic mean when random mating is assumed.

$$Ne = \frac{t}{\frac{1}{N1} + \frac{1}{N2} + \cdots + \frac{1}{Nt}},$$

where N is the population size and t is the number of generations considered in the analysis. A rare crash in size, which frequently occurs through drought or disease, clearly has an important effect on effective numbers (Fig. 4.3). This can be demonstrated by the example of a random mating population which normally maintains an effective size of 100 and then drops for one generation to 6. Then over a ten-generation interval, the effective population size accounts for only 39.

All these factors and additional ones like the actual size of the mating group in space (neighbourhood size) and dispersal rates among neighbourhoods influence the effective population size independently. Reed et al. (1986) present an example of a population estimate for the elk (*Cervus elaphus*). Because of overlapping generations, a skewed sex ratio in the breeding population, and a long period before their first reproduction, there must be at least 426 individuals in an unhunted population to maintain an effective population size (Ne) of 50 elks. Their calculation, however, does not take into account variability in

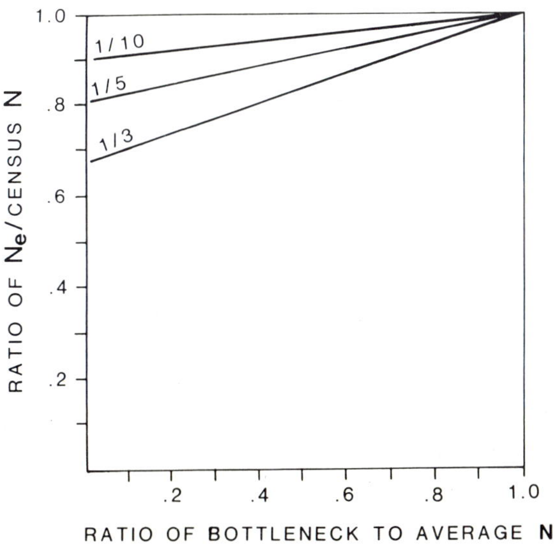

Fig. 4.3. The relationship between effective population size and census size as a function of how census numbers vary over time. Bottlenecks (size minima) can have disproportionate effects but only if they are very small or occur frequently. The *lines* indicate bottlenecks that occur from once in 3 to once in 10 generations. (After Chepko-Sade et al. 1987)

reproductive success among individuals; it assumes a stable age distribution and population size. Also, if mating is not random, more breeding pairs are required to maintain the given effective population size of 50. This is because the relative overproduction by one pair increases the ratio of their genes in a population (Reed et al. 1986). If this occurs for several generations, the population will lose genetic variability through fixation of the alleles of the "overproducers". This is also the reason why polygamous species require a greater number of individuals than monogamous species (all other things being equal) to maintain a desired effective population level (Reed et al. 1986).

Currently, there is no commonly accepted method for combining all these different factors into a single estimate. Perhaps the easiest method might be to calculate a final Ne by using the formula associated with each factor sequentially, i.e., by using the Ne derived from the first factor as the initial estimate of N in the next formula and so on until the contributions of all factors have been included (Chepko-Sade et al. 1987).

However, a single instance of close inbreeding, for example, the founding of a population with a small number of individuals, does not necessarily mean that the descendants will be highly inbred. Relatively few unrelated founders (20 to 30) can provide an astonishingly significant sample of the average diversity of the pertinent gene pool (Senner 1980; Fig. 4.4). Unfortunately, relict populations are rarely "ideal populations" and may already have experienced some inbreeding. To avoid further loss of genetic variability and to increase the chance of survival these populations should be expanded as rapidly as possible (Senner 1980; Ralls and Ballou 1983; Foose et al. 1987). To a great extent, preservation of genetic

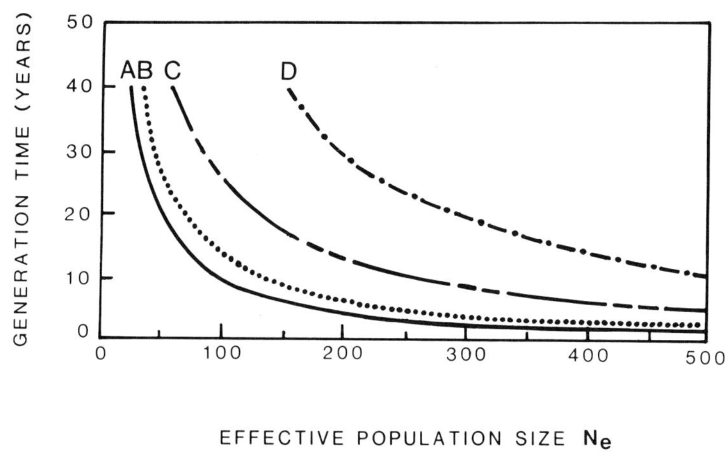

Fig. 4.4. Effective population size required to preserve 90% of original genetic diversity for 200 years if population is established by various numbers of founders, assuming immediate expansion of founder groups with no loss of diversity. *A* No founder effect; *B* 20 founders; *C* 8 founders; *D* 6 founders. (After Foose et al. 1987 from a graph prepared by M. Soule and J. Ballou)

diversity is a function of the size of the population being maintained or protected. The smaller the population, the faster genetic diversity is lost, principally because of random genetic drift (Foose et al. 1987; Fig. 4.1). Genetic variation is necessary for evolution to occur and provides a basis for populations to adapt to changes in the environment (Reed et al. 1986; Lande and Barrowclough 1987). Populations on the verge of extinction, which may be virtually devoid of genetic variability (and perhaps with a reduced fitness as a result of inbreeding), should not be considered hopeless because, if the population regains moderate or large numbers, genetic variation can be regenerated by mutation, thus restoring the potential for adaptive evolution (Lande and Barrowclough 1987). An effective population size (Ne) of 50 is considered to be sufficient to avoid loss of variability due to inbreeding. However, a much larger effective population size is required to neutralize the effects of genetic drift (Franklin 1980). Genetic drift is referred to as the variability that is lost by chance, e.g., individuals with certain genetic traits die before reproducing (Frankel and Soule 1981). As a first approximation, a genetically effective population size of 500 is considered to be satisfactory for the accommodation of continuing evolution (Lande 1976; Franklin 1980). On theoretical grounds, the drift in populations of this size is not great (Fig. 4.1) and should be countered by natural selection with little selective death (Franklin 1980). The loss of genetic variation should be approximately balanced by the gain from mutation. Again, it has to be stressed that the actual number of individuals in the population, for the reasons given above, may have to be several times greater than 500 (Frankel and Soule 1981).

4.3 Demographic and Spatial Considerations

So far I have only dealt with numbers of animals, but in natural populations this transforms into required living space. Space is probably the most important and limiting factor involved in maintaining viable natural populations of mammals. Minimum viable populations can be theoretically maintained indefinitely in captivity but it is not desired in the long term, since natural selection is missing.

The extinction history of mammal populations is strongly interconnected with the theory of island biogeography (MacArthur and Wilson 1967). Islands that are separated from the mainland by land bridges go through a process of "faunal relaxation". This means that, after isolation, extinction processes predominate until a faunal equilibrium is reached. The smaller the island, the greater the number of extinctions until an area-specific species number is reached (Terborgh and Winter 1980). The same extinction pattern can be observed in western North American Parks. After their creation (less than 100 years ago) extinctions occurred in relation to area size (Newmark 1987; Fig. 4.5). Only the largest park assemblage (20, 736 km^2). still contains an intact historical mammalian faunal assemblage. However, the theory of island biogeography fails to increase the quality of the species concerned. Each species in the models is treated equally, but in reality we can observe certain trends. Some species have

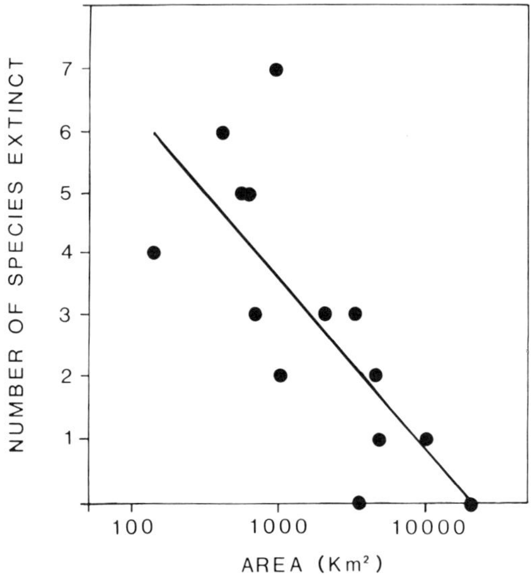

Fig. 4.5. Relationship between number of natural post-establishment extinctions and park area in 14 western North American national parks. (After Newmark 1987)

a greater chance of survival and others are always the ones that suffer extinction first. The reason for this is that different species have different minimum area requirements which depend on body size and feeding type. In a study of shrews on islands of Lake Inari in northern Finland, Peltonen et al. (1989) did not encounter any *Sorex caecutiens* on islands smaller than 3 ha. The lower size limit for the occurrence of the marsupial *Potoros tridactylos* in a sample of woodlots ranging in size from 0.3 to 82 ha was sightly less than 10 ha (Bennett 1987). Nevertheless, the occurrence of individuals of a species on an island or in a habitat patch does not necessarily mean that they carry viable populations. The consequences will be discussed further on. The patterns that underlie the processes of selective extinctions on islands are presented in the concept of "nested subsets" of species assemblages in land-bridge, insular, faunal assemblages (Patterson and Atmar 1986; Patterson 1987, 1990).

Newmark (1986, cited in Soule 1987c) has shown that the most consistent predictor of persistence of mammalian species in western North American national parks is the estimated population size. Since carnivores have smaller population densities than herbivores, and large mammals of the same trophic level have smaller densities than small ones, the largest mammalian carnivores are expected to go first, followed by the largest herbivores. The persistence time depends on the size of the reserve and, therefore, ultimately on population size. Such an extinction pattern can be observed on the islands of the Sunda Shelf, Southeast Asia, which became separated from the mainland due to rising sea level when the last glaciation terminated (Heaney 1984). Among carnivores, the largest species show high extinction probabilities on medium and small islands; tigers, leopards, bears, badgers, otters, and wild dogs are nearly always absent on any but the largest islands, whereas the smaller (and more omnivorous) viverrids show relatively lower extinction probabilities. Large herbivores show a similar pattern: elephants, wild cattle, rhinoceroses, and tapirs are absent from any but the three largest islands, whereas the medium-sized Malayan sambar and the bearded pig are more widespread, and small mouse deer are abundantly represented on small islands (Heaney 1984).

Terborgh and Winter (1980) also state that rarity emerges as a reasonably good predictor of vulnerability to extinction. Nearly all rare species conform to one of the three following patterns: (1) species that are near the limits of their geographic ranges or habitat spectrum; (2) species that are constitutively rare, that is, they exist at low population densities wherever they occur (e.g., large predators); and (3) species that specialize on certain types of patchily distributed habitat (Terborgh and Winter 1980). Newmark (1986, cited in Soule 1987c) found that the initial sizes for most surviving populations of mammals in western North American parks fall within the range of 100 to a few thousand. Surviving lagomorph populations are about two orders of magnitude higher, a probable effect of the legendary outbreak-crash populations of rabbits and hares (Table 4.1). Persistency over longer periods of time (more than 100 years) would probably require larger numbers (Soule 1987c).

Belovsky (1987) estimates that the largest mammalian carnivores can be expected to persist in 0–22% of the current parks, but no preservation area is large enough to guarantee persistence for 1000 years. The future for natural populations of the largest herbivores is brighter, since 4–100% of the reserves should permit persistence for 100 years and 0–22% will permit persistence for 100 years, under the assumption that major climatic variations do not occur. Peters and Realson (1984) estimated the size of now extinct populations on

Table 4.1. Comparisons of estimated, initial, and median sizes for extinct populations of different groups of mammals with the estimated, initial median sizes for surviving populations in 24 western North American national parks and park assemblges (rearranged after Soulé 1987c, based on data from Newmark 1986)

Group	S/Ex[a]	(n)	Median population size	95% confidence interval	Ratio median (S:Ex)
Lagomorphs	S	60	70 889	34 720–173 150	21.6
	Ex	9	3276	702–56 952	
Small carnivores	S	127	1203	908–1704	4.7
	Ex	26	256	122–880	
Artiodactyls	S	88	792	429–1504	3.3
	Es	7	241	3–1273	
Large carnivores	S	153	108	70–146	4.5
	Ex	16	24	14–68	

[a] S = Extant populations; Ex = extinct populations.

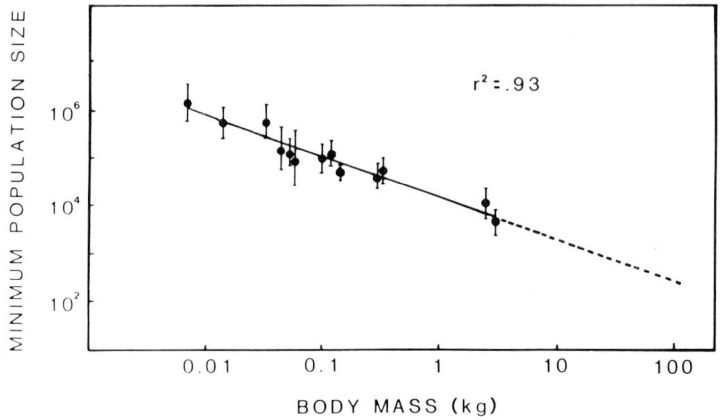

Fig. 4.6. Effects of body size on an estimate of the minimum size for a sustainable mammalian population. *Vertical bars* indicate 95% confidence limits around the mean estimate of now extinct populations of boreal mammals thought to have inhabited mountaintop islands in the Great Central Basin. A self-sustaining population should probably exceed that calculated from the regression through the mean values: $N = 16\,000\,W^{-0.85}$. (After Peters and Realson 1984)

mountaintops that became isolated about 10000 years ago after the end of the last glaciation. The data on possible occurrences of now extinct taxa on mountaintop islands were taken from Brown (1971). Based on available space, and the average population density of the species reported in the literature, they came up with a geometric mean of the calculated size of all 124 now extinct populations of 108000 animals and the 95% confidence limits around the mean were 75000 and 156000. Such calculations provide only an order of magnitude estimate since it was assumed that the entire island (mountaintop) was habitable. However, it shows that for the long-term survival of self-sufficient mammal populations extremely large population sizes may be required. The estimated minimum population sizes also seem to depend on body size (Peters and Realson 1984). To illustrate this effect the calculated mean sizes of extinct populations were plotted against mean body mass (Fig. 4.6). The curve suggests that a population of at least 16000 animals is required to preserve a 1-kg animal. This minimum size falls rapidly as body mass increases. Self-sustaining populations of 100-kg mammals should consist of more than 320 individuals. Above a body mass of 25 kg, the minimum sustainable population calculated by extrapolating the curve given by the authors falls below 1000. They themselves consider these data to have some value in estimating the minimum population density which is probably required for long-term population survival. However, they omit that much more confirmation would be required before any implications of the results can be given credence. Naturalists have long known that most species exist in subdivided populations (demes) which live in distinct habitat patches or islands (Chepko-Sade et al. 1987). They form a so-called metapopulation.

Levins (1971) introduced this concept which he described as follows: any real species is a population of local populations which are established by colonists, survive for a while, send out migrants, and eventually disappear. The persistence of a species in a region depends on the rate of colonization successfully balancing the local extinction rate. However, if every population has a finite life expectancy, the absolute distinction between stable and unstable communities breaks down into a continuum of extinction probabilities. Because of the transient character of local populations, the concept fits well with the mosaic-cycle concept of ecosystems (Remmert 1991).

Rates of gene flow between subpopulations depend on the degree of isolation, normally mediated by physical distance and the character of intervening habitat. Patch extinction depends on the genetics and demography of the population and environmental forces that act within patches. Patch recolonization depends on whether neighbouring patches are occupied, how big they are and how near they are. Generally, larger islands produce more migrants, the probability of immigration falls exponentially with distance and the probability of extinction is inversely related to the area or carrying capacity of the patch (Gilpin 1987). A change in any of these factors may lead to the extinction of entire metapopulations. Metapopulation survival requires a mean extinction rate that is less than the immigration rate (Levins 1970).

A population is genetically isolated if less than one effective migrant enters a population per generation (Lande and Barrowclough 1987); to be "effective" a migrant must actually breed. If patch populations turn over, each patch population will have gone through a recent genetic "bottleneck" during the colonization period. Such "bottlenecks" will greatly exacerbate inbreeding (Gilpin 1987; Fig. 4.3).

Clearly, a small inbred patch population cannot have much genetic variance as was shown for the wolves of Isle Royal (Wayne et al. 1991) and the lions of the Ngorongoro Crater (Packer et al. 1991). Yet somewhat paradoxically, a system of patch populations considered at the regional level may have more genetic variation than would a single large population of the same total size (Gilpin 1987). There are two reasons for this. First, local populations may fix different alleles. Second, the selection pressures may be different in different environmental patches. In any case, the genetic variation between patches can be very useful in protecting the species from extinction (Gilpin 1987). Different patches might contain the genetic variants that could survive strong selection pressures, such as a disease. When there is a frequent turnover of patch populations, each local population will have derived all of its genetic variation from a recent pair of founders, and the effects of local drift and past selection will be overwhelmed. The metapopulation will have a very homogeneous structure of low genetic variance, which has apparently happened with the cheetah (O'Brian et al. 1983) and may have occurred as well for the black-footed ferret (Gilpin 1987). Examples for a metapopulation of mammals are small mammals that live on lake islands (Peltonen et al. 1989) or isolated woodlots of virgin forest (Bennett 1987), but also neighbourhood groups of killer whales, which seem to have almost no genetic exchange, other marine mammals that display long-term fidelity to particular breeding sites (Harwood and Hall 1990) or black-footed ferret demes in prairie dog colonies (Brussard and Gilpin 1989). From DNA fingerprinting we know that pods of killer whales and colonies of naked mole rats are highly inbred (Faulkes et al. 1990; Harwood and Hall 1990). However, so far, we do not know the implications that this might have for the possible survival of individual demes. Nevertheless, with the presence of strong selection pressure, inbreeding should not lead to a manifestation of deleterious genes, when weak animals do not survive until adulthood or are socially excluded from reproduction.

Nowadays our conservation areas have become increasingly isolated so that we can treat them as true islands without natural immigration. Many areas where mammals became extinct in the past will need to be restocked artificially since there is no animal pool nearby for natural repopulation. Only by the translocation of animals or genetic material in the form of sperm or embryos can a gene flow between subpopulations be maintained for many species of large mammals as a means of increasing effective population sizes. Immigration of unrelated individuals into an inbred population reduces the level of inbreeding dramatically (Soule 1980; Salwasser et al. 1984; Lande and Barrowclough 1987). Due to artificial immigration, small, physically isolated populations can maintain a high level of heterozygosity as shown by feral goats on the island of Montecristo, Italy (Randi et al. 1990).

4.4 Environmental and Catastrophic Considerations

Environmental and catastrophic uncertainty may also require very large population sizes for a high level of security in the mid to long term (e.g., 95% probability of persistence for 100 or 1000 years; Belovsky 1987; Ewens et al. 1987; Goodman 1987; Schaffer 1987). The persistence times predicted for populations subject to substantial environmental variation can be distressingly short, and these persistence times are not dramatically increased by modest increases in the maximum population size (Goodman 1987). For some populations which are sensitive to environmental fluctuations, the minimum reserve size necessary to achieve a reasonably long persistence time is extremely large, on demographic grounds alone (Goodman 1987). Serious attention must be given to field programs which estimate the relevant parameters of environmental variation—in particular, the magnitude of variance that environmental variation causes in realized population growth rates, and the degree of correlation in that variation over space and time (Goodman 1987).

To cope with large-scale environmental catastrophes, like drought or extreme winters, several populations must be preserved or established, provided that environmental variation in the separate areas is at least partially independent. A "local catastrophe" to a single patch population may not be a catastrophe to the entire metapopulation. The patch can be maintained as living space for the species, provided there is at least a small rate of natural or managed recolonization into areas which experience local extinctions.

The general, negative genetic effects of splitting up a single population into a metapopulation with semi-isolated subunits may be outweighed by the positive effects of security against localized environmental catastrophes and diseases. Nevertheless, large-scale perturbations, like volcanism, regional pollution, large fires, floods, etc., can affect the metapopulation as a unit.

4.5 Conclusions

As shown above from a genetic perspective, an effective size of 500 will usually translate into a census size of several times this number, when taking into account all of the determining factors including age structure, sex ratio, variance in family size, and fluctuations in population size over time (Soule 1987c; Lande and Barrowclough 1987).

A similar decrease in the relation of the census population to the effective population size holds true for populations in fluctuating environments (Goodman 1987) or for species that are subjected to wide swings in population size, such as lagomorphs (Franklin 1980). Both face an increased probability of chance extinctions (Belovsky 1987; Goodman 1987). In comparison to stable populations, these populations require MVPs that are therefore much larger, often by orders of magnitude (Soule 1987c).

Genetic variation in a founder population rapidly approaches that in a wild source population once the effective size of the founder population exceeds 25 (Ralls and Ballou 1983; Foose et al. 1987). To go once through a "bottleneck" does not necessarily mean that a great percentage of heterozygosity is lost (Chepko-Sade et al. 1987), but the population needs to be expanded as rapidly as possible (Senner 1980). The maintenance population size is much more critical in the preservation of heterozygosity than the actual number of founders (Senner 1980). With equally effective population sizes, species with longer generation times will lose less genetic variability in a given period of time than species with shorter generation times, because, if population size remains constant, small populations lose genetic variability only when they reproduce. One practical implication of such calculations is that final population sizes can be smaller for larger animals with long generation times, such as elephants, than for small ones with short generation times, such as many rodents (Ralls and Ballou 1986; Fig. 4.7). Nevertheless, we still lack empirical tests to determine the minimum size of a viable population (Thomas 1990; Saether and Jonsson 1991) and an integrated predictive model to determine the potential longevity of populations in relation to their size and structure. According to Schaffer (1987), extinction dynamics also remains inadequately understood because:

1. No model is comprehensive enough to incorporate all types of environmental and demographic uncertainties in addition to different life history traits, and

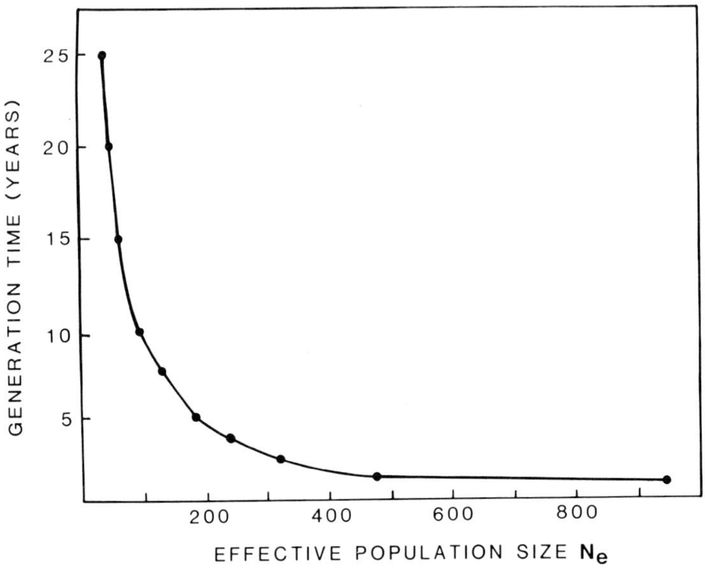

Fig. 4.7. Effective population size necessary to preserve 90% of original genetic diversity for 200 years for taxa with various generation times. (After Foose et al. 1987)

2. Even for the models now available, the necessary data for actual application are generally lacking and are both very costly and time-consuming to gather (Schaffer 1987). Initial attempts to estimate MVPs and minimum area requirements (MARs) for unsubdivided populations of mammals provide a range of population sizes of hundreds to millions and areas up to millions of square kilometers for high-level, mid- to long-term security (i.e., 95% probability of persistance for 100 or 1000 years, respectively). The range is set by population growth rates, their variability and population density (Belovsky 1987; Schaffer 1987). It is clear, based on crude estimates of extinction probabilities, that the size and number of the current nature reserves are probably insufficient to provide a high level of long-term security for at least some mammalian species, especially those that are large or rare (Schaffer 1987).

Nevertheless, from the models and the limited evidence we have from real populations, we can conclude that there is no single value or "magic number" for the population size that has universal validity (Soule 1987b). A viability analysis must be performed for each case, because each case is different (Soule 1987c). Not only is each situation unique, but the acceptable level of risk determines the numbers, densities, and distribution in space of the MVP (Schoenwald-Cox 1983).

Acknowledgements. I thank Ursula Korn for helpful comments on various drafts of the manuscript.

References

Allendorf FW, Leary RR (1986) Heterozygosity and fitness in natural populations of animals. In: Soule ME (ed) Conservation biology. The science of scarcity and diversity. Sinauer Associates, Sunderland, pp 57–76

Belovsky GE (1987) Extinction models and mammalian persistence. In: Soule ME (ed) Viable populations for conservation. University Press, Cambridge, pp 35–57

Bennett AF (1987) Conservation of mammals within fragmented forest environment: the contributions of insular biogeography and autecology. In: Saunders DA, Arnold GW, Burbidge AA, and Hopkins AJM (eds) Nature conservation: the role of remnants of native vegetation. Beatty and Sons, Surrey, in association with CSIRO and CALM, pp 41–52

Brown JH (1971) Mammals on mountaintops: nonequilibrium insular biogeography. Am Nat 105: 467–478

Brussard PF, Gilpin ME (1989) Demographic and genetic problems of small populations. In: Seal US, Thorno ET, Bogan MA, Anderson SH (eds) Conservation biology and the black-footed ferret. Yale University Press, New Haven, pp 37–48

Chepko-Sade BD, Shields WM, Berger J, Halpin ZT, Jones WT, Rogers LL, Rood JP, Smith AT (1987) The effects of dispersal and social structure on effective population size. In: Chepko-Sade BD, Halpin ZT (eds) Mammalian dispersal patterns. The effects of social structure on population genetics. University Press, Chicago, pp 287–321

Crow JF, Kimura M (1970) An introduction to population genetics theory. Burgess, Minneapolis

Crow JF, Morton NE (1955) The measurement of gene frequency drift in small populations. Evolution 9: 202–214

De Bois H, Dhondt AA, Van Puijenbroek B (1990) Effects of inbreeding on juvenile survival of the okapi *Okapia johnstoni* in captivity. Biol Conserv 54: 147–155

Ewens WJ, Brockwell PJ, Gani JM, Resnick SI (1987) Minimum viable population size in the presence of catastrophes. In: Soule ME (ed) Viable populations for conservation. University Press, Cambridge, pp 59–68

Faulkes CG, Abbott DH, Mellor AL (1990) Investigation of genetic diversity in wild colonies of naked mole-rats (*Heterocephalus glaber*) by DNA-fingerprinting. J Zool (Lond) 221: 87–97

Foose TJ, Seal US, Flesness NR (1987) Captive propagation as a component of conservation strategies for endangered primates. In: Marsh CW, Mittermaier RA (eds) Primate conservation in the tropical rain forest. Monographs in primatology, vol 9, Allan R Liss, New York, pp 263–299

Frankel OH (1983) The place of management in conservation. In: Schoenwald-Cox CM, Chambers SM, MacBryde B, Thomas L (eds) Genetics and conservation. Benjamin/Cummings, Menlo Park, pp 1–14

Frankel OH, Soule ME (1981) Conservation and evolution. Cambridge University Press, London, 327 pp

Franklin IR (1980) Evolutionary change in small populations. In: Soule ME, Wilcox BA (eds) Conservation biology: an evolutionary-ecological perspective. Sinauer, Sunderland, pp 135–149

Futuyma D (1986) Evolutionary biology, 2nd edn. Sinauer, Sunderland

Gilpin ME (1987) Spatial structure and population vulnerability. In: Soule ME (ed) Viable populations for conservation. University Press, Cambridge, pp 125–139

Goodman D (1987) The demography of chance extinction. In: Soule ME (ed) Viable populations for conservation. University Press, Cambridge, pp 11–34

Harwood J, Hall A (1990) Mass mortality in marine mammals: its implications for population dynamics and genetics. TREE 5: 254–257

Heaney LR (1984) Mammalian species richness on the Sunda Shelf, Southeast Asia. Oecologia (Berl) 61: 11–17

Lande R (1976) The maintenance of genetic variability by mutation in a polygenic character with linked loci. Genet Res (Camb) 26: 221–235

Lande R (1979) Effective deme sizes during long-term evolution estimated from rate of chromosomal rearrangement. Evolution 33: 234–251

Lande R, Barrowclough GF (1987) Effective population size, genetic variation, and their use in population management. In: Soule ME (ed) Viable populations for conservation. University Press. Cambridge pp 87–123
Levins R (1970) Extinctions. In: Gerstenhaber M. (ed) Some mathematical questions in biology, vol II. American Mathematical Society, Providence, RI pp 77–107
Levins R (1971) Evolution in changing environments, 2nd edn. University Press, Princeton
MacArthur RH, Wilson EO (1967) The theory of island biogeography. Princeton, New Jersey, 203 pp
Neuhäuser P (1991) Populationsmanagement und Artenschutz aus genetisch-demographischer Sicht. Biol Zentralbl 110: 87–113
Newmark WD (1986) Mammalian richness, colonization and extinction in western North American national parks. PhD Dissertation, University of Michigan, Ann Arbor (cited after Soulé 1987c)
Newmark WD (1987) A land-bridge island perspective on mammalian extinctions in western North American parks. Nature 325: 430–432
O'Brien SJ, Wildt DE, Goldman D, Merril CR, Bush M (1983) The cheetah is depauperate in genetic variation. Science 221: 459–462
Packer C, Pusey AE, Rowley H, Gilbert DA, Martenson J, O'Brien SJ (1991) Case study of a population bottleneck: lions of the Ngorongoro Crater. Conserv Biol 5: 219–230
Patterson BD (1987) The principle of nested subsets and its implications for biological conservation. Conserv Biol 1: 323–334
Patterson BD (1990) On the temporal development of nested subset patterns of species composition. Oikos 59: 330–342
Patterson BD, Atmar W (1986) Nested subsets and the structure of insular mammalian faunas and archipelagos. Biol J Linn Soc 28: 65–82
Peltonen A, Peltonen S, Vilpas P, Beloff A (1989) Distributional ecology of shrews in three archipelagos in Finland. Ann Zool Fenn 26: 381–387
Peters RH, Realson JV (1984) Relations between individual size and mammalian population density. Am Nat 124: 498–517
Ralls K, Ballou J (1983) Extinction: lessons from zoos. In: Schoenwald-Cow CM, Chambers SM, MacBryde B, Thomas L (eds) Genetics and conservation. Benjamin/Cummings, Menlo Park, pp 164–184
Randi E, Tosi G, Toso S, Lorenzini R, Fusco G (1990) Genetic variability and conservation problems in Alpine ibex, domestic and feral goat populations (genus *Capra*). Z. Säugetierk 55: 413–420
Reed JM, Doerr PD, Walters JR (1986) Determining minimum viable population sizes for birds and mammals. Wildl Soc Bull 14: 255–261
Remmert H (ed) (1991) The mosaic-cycle concept of ecosystems. Ecological studies 85. Springer, Berlin Heidelberg New York
Salwasser H, Mealey SP, Johnson K (1984) Wildlife population viability: a question of risk. Trans N Am Wildl Resour Conf 49: 421–439
Schaffer M (1987) Minimum viable populations: coping with uncertainty. In: Soule ME (ed) Viable populations for conservation. University Press, Cambridge, pp 69–86
Shaffeer ML, Samson FB (1985) Population size and extinction: a note on determining critical population sizes. Am Nat 125: 144–152
Schonewald-Cox CM (1983) Guidelines to management: a beginning attempt. In: Schonewald-Cox CM, Chambers SM, MacBryde B, Thomas L (eds) Genetics and conservation. Benjamin Cummings, Menlo Park, pp 414–445
Senner JW (1980) Inbreeding depression and the survival of zoo populations. In: Soule ME, Wilcox BA (eds) Conservation biology: an evolutionary-ecological perspective. Sinauer, Sunderland, pp 209–224
Shields WM (1982) Philopatry, inbreeding and the evolution of sex. State University of New York Press, Albany
Slatis HM (1960) An analysis of inbreeding in the European bison. Genetics 45: 275–288
Soule ME (1980) Thresholds for survival: maintaining fitness and evolutionary potential. In: Soule ME, Wilcox BA (eds) Conservation biology: an evolutionary-ecological perspective. Sinauer, Sunderland, pp 151–169

Soule ME (ed) (1987a) Viable populations for conservation. University Press, Cambridge, 189 pp

Soule ME (1987b) Introduction. In: Soule ME (ed) Viable populations for conservation. University Press, Cambridge, pp 1–10

Soule ME (1987c) Where do we go from here? In: Soule ME (ed) Viable populations for conservation. University Press, Cambridge, pp 175–183

Templeton AR (1987) Inference on natural population structure from genetic studies on captive mammalian populations. In: Chepko-Sade BD, Halpin ZT (eds) Mammalian dispersal patterns. The effects of social structure on population genetics. University Press, Chicago, pp 257–272

Terborgh J, Winter B (1980) Some causes of extinction. In: Soule ME, Wilcox BA (eds) Conservation biology: an evolutionary-ecological perspective. Sinauer, Sunderland, pp 119–133

Thomas CD (1990) What do real populations tell us about minimum viable population size? Conserv Biol 4: 324–327

Wayne RK, Gilbert DA, Lehman N, Hansen K, Eisenhawer A, Girman D, Peterson RO, Mech LD, Gogan PJP, Seal US, Krumenaker RJ (1991) Conservation genetics of the endangered Isle Royal gray wolf. Conserv Biol 5: 41–51

5 On Micropopulations in Fish and Their Effects on Differentiation and Speciation

W. VILLWOCK

5.1 Introduction

It is a surprising fact that there are only a few exactly investigated and therefore really well-known examples of the effects of small population size in fish on differentiation (segregation) and speciation. This is the fact, although it has long been known that a small number of specimens will do as "starting capital" for species distribution and development, very often artificially caused by man (e.g. three pairs of bisam, *Ondatra zibethicus*, introduced in 1905 from Canada to a farm in Bohemia/former Czechoslovakia, from where they escaped, reproduced and distributed throughout central Europe within an unbelievably short period of less than 50 years).

However, these examples do not cover the question under discussion, because they proceed from a given species and do not result in one or more new one(s), even after being introduced to strange ecological environments. Examples belonging to this kind of developmental consequences are legion; in fin-fish alone some thousand different ones are documented (Rosenthal 1993). A few of them and their consequences on the native fauna were recently reported in detail by the author (Villwock 1993).

The present contribution on "small populations" will reflect on those processes by which remarkable interactions may be illustrated between population size and differentiation of certain characters (used in its broadest sense) and speciation mechanisms.

5.2 Micropopulations in Anatolian *Aphanius* ssp. (Fam.: Cyprinodontidae)

As far as is known, Kosswig (1956) was the first to report expressis verbis on (macro- and) micropopulations in fish, exemplified by endemic Anatolian cyprinodontids of the bittersalt Lake Aci-Göl, which is located in the southwest of Turkey.

During the summer dry season, the shallow lake is transformed to an enormous degree into crystalline sodium sulfate. Fish life is then restricted to small brooks and springs around the border of the former lake. During the winter the central part of the lake is transformed into a hypersaline salt swamp

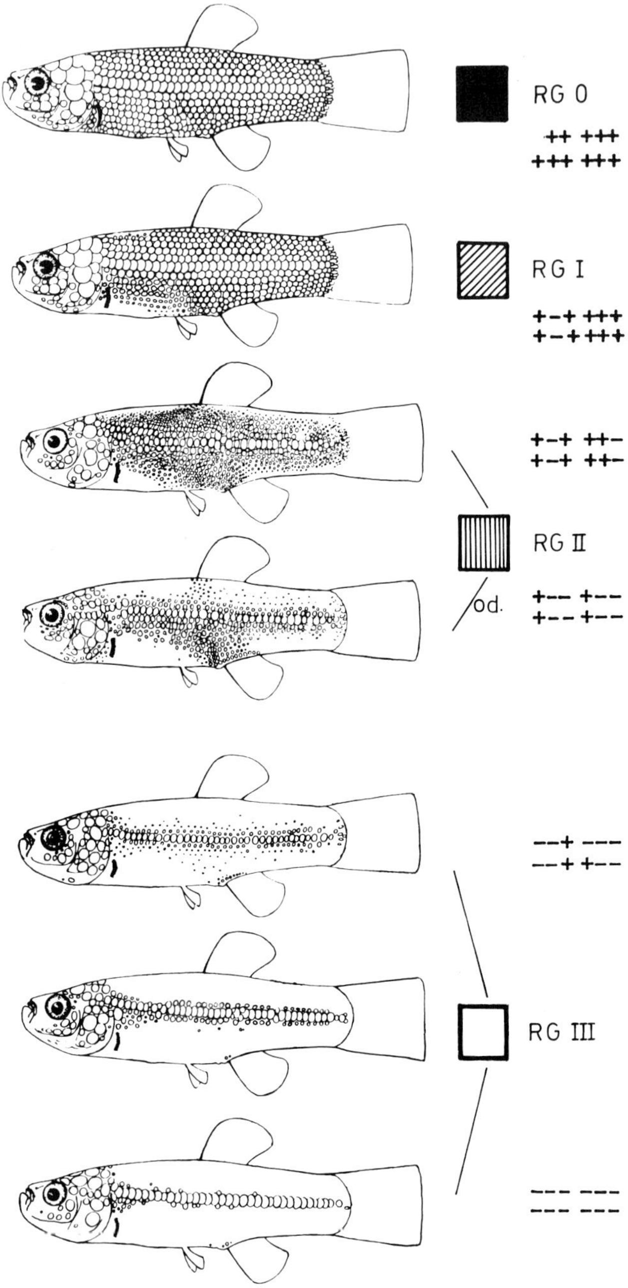

Fig. 5.1

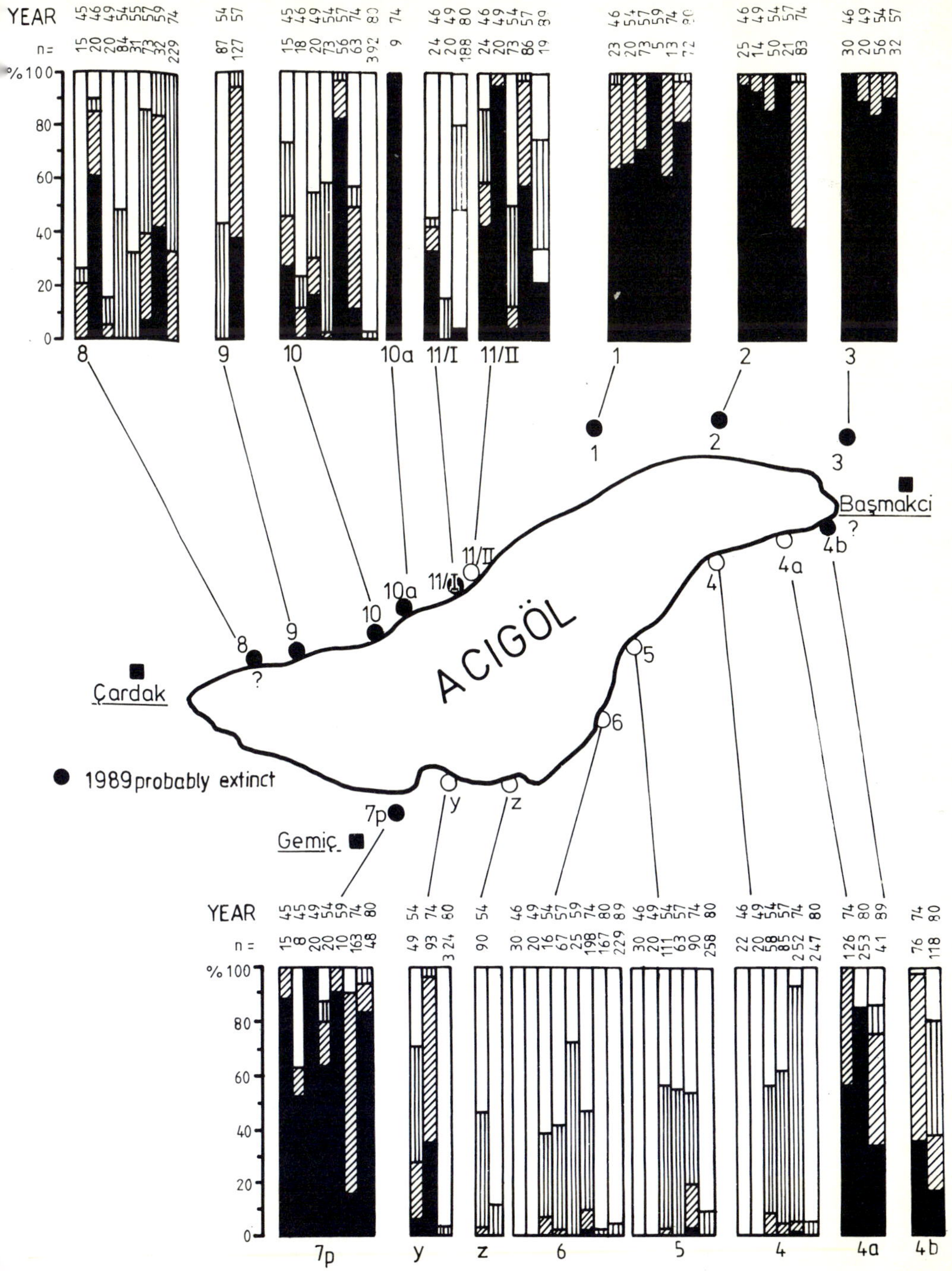

Fig. 5.1. Aci-Göl with population graphs and degree of scale formation p.a. *RG* Reduction degree. (drawing M. Hänel, Hamburg)

with salt concentrations too high to allow migration of fish from one lake side to the other.

It is highly interesting that the entire group of populations of the Aci-Göl cyprinodonts—described as descendants of an old panmictic "hybrid swarm" (Kosswig 1956; Villwock 1958)—are so very different from each other that an ichthyologist without sufficient knowledge of the special biota or a large collection of material is misled to describe those forms at least as two species of two distinct genera (Aksiray 1948; Aksiray and Villwock 1962; Villwock 1963). As far as body shape, indices and scale formation are concerned, three groups of populations can be distinguished from one another:

Populations 1–3 (Fig. 5.1) occur in small brooks of slowly running freshwater; six populations (nos. 4a–7y/z) live the whole year round in immediate contact with the more or less highly concentrated bittersalt water of the lake, while the remaining populations (nos. 7p–11/II) exist under neither of these extreme environmental conditions, but under changing ones. Depending on the amount of water produced by their very strong fountains, as well as the effectiveness of rainfall and melting snow during springtime, they are exposed to changing conditions; sometimes (and often for a longer period) freshwater conditions predominate, occasionally sodium sulphate water predominates.

Corresponding to these different ecological factors there are three groups of populations, consisting of very few hundreds to many ten thousands of specimens.

Correlated with freshwater, the population structure shows a more or less normal scale formation and *Aphanius*-like body shape. The following population assembly reflects the extreme bittersalt conditions by almost complete scale reduction, concerning size, shape and number of scales as well as the degree of scaliness. Additionally, these specimens show body relations different from those of *Aphanius*, for which reasons they were said to belong to a separate genus, *Anatolichthys* (Kosswig and Sözer 1945). The specimens of populations living under changing water conditions show "mixed" characters with respect to scaliness and general characters (see Fig. 5.1).

In spite of all these differences, the members of the different populations proved to be interfertile in their hybrids. From this, the conclusion must be drawn that even microgeographic separation will not automatically lead to valid speciation, even if the external characters differ as far as they do in this example—and for which Mendelian inheritance can be shown (Villwock 1958; Aksiray and Villwock 1962; Villwock 1963).

It is not the intention to bring up the whole story here again. However, in the context of this chapter, the already mentioned fact may be of interest that the characters in discussion are under polygenic control and of Mendelian heredity. Moreover, they are undoubtedly correlated to selectively acting physiological factors, triggered by predominant water conditions, while the degree of their expressivity depends on population size and the distribution of the genes or gene complexes individually involved. The given genetic background may have derived from the above-mentioned ancestral panmictic macropopulation in which the heterogeneity concerning different types of scale formation or scale reduction came into action by substitution of genes and transfer of gene function.

Such a “mixed” macropopulation still exists in Lake Burdur, next to Lake Aci-Göl, which may serve as a model for selectively acting isolating mechanisms when transferred to micropopulations by historic changes, e.g. of climate: it is generally accepted that the different phenomena of Aci-Göl *Aphanius* came stepwise into existence, when the old Aci-Göl changed its character from a true lake to a more or less bittersalt swamp with a surrounding chain of springs, ponds and brooks, since in postglacial times the climate became more and more arid (Kosswig 1956; Villwock l.c.).

The same is true in a corresponding sense with the desert water remains in the southwestern United States, especially in the Death Valley system, Nevada, and in northern Mexico: their inhabitants, representatives of another cyprinodont genus, *Cyprinodon* ssp., reflect in a startling manner very similar adaptations to their predominant water conditions to those of the Old World *Aphanius*, even with the same consequences to the display of speciation phenomena (Corcillius 1979).

5.3 Intralacustrine Segregation Phenomena in East African Cichlids (Cichlidae)

The second famous group of fishes interesting in the context of this contribution belong to the family of Cichlidae, which are well known as coloured perches; 200 to 300 are endemic to the East Africa rift valley lakes.

Best investigated among them are the *Haplochromis* ssp. and their relatives from Lake Victoria (and others), summarized as “haplochromines” (Greenwood 1964, 1973). Another famous species flock is that of *Tropheus* ssp. and about ten related genera (McElroy and Kornfield 1990), the so-called Mbuna, from Lake Tanganyika, Lake Nabugabo, Uganda, and other lakes of the East African rift valley system.

The phenomena in discussion are described as “explosive speciation”. The term “...was coined to cover the phenomenon of very rapid evolution of numerous closely related species in a circumscribed geographical area” (Greenwood 1964, p. 256). The phenomenon itself has long been discussed as to its way allopatric (as usual) or sympatric—or to use an other term: intralacustrine (Kosswig 1947; Trewavas 1947, 1948; Fryer 1959).

There is no doubt that the latter mode had taken place in the very old and hugest fresh-water lake in the world, Lake Baikal in Siberia, Russia (Kohzov 1963).

Intralacustrine speciation is also accepted for many organism groups of Lake Ohrid from Macedonia-Skopje/Alban, while the *Salmo letnica* population (fam. Salmonidae) begins to divide into four distinct species in statu nascendi within the border of Lake Ohrid by forming four distinct reproduction units (Stankovic 1953; Kosswig 1963 a,b).

Another example which is suggested as showing intralacustrine species formation is that of Lake Titicaca, situated in the Altiplano of Peru/Bolivia, with its rich and endemic *Orestias* ssp. (Fam. Cyprinodontidae) (Villwock 1986).

The main objection to the assumption of intralacustrine speciation is the insufficiently explained interruption of panmixis between dividing species in statu nascendi. However, according to the author's conviction, the whole problem seems to be a statistical one: if intralacustrine radiation acts on the background of enough different habitat niches with a corresponding adaptation to feeding and/or reproduction strategies—both associated with "fitting" ethological, i.e. behavioural "structures"—the result should soon be differently developing populations of a given species.

Even if there were some "intervening" specimens (without specific adaptations) capable of reproducing with some of their former conspecifics, the isolation of both separating groups will not be stopped or at least reversibly influenced by this occasional interbreeding: the mass of specimens of the processing populations will mainly reproduce with specimens of the same kind of differently ongoing specialization. "Unspecialized" specimens will, sooner or later, no longer be accepted by sex partners of the dividing groups, i.e. by non-fitting mating patterns (in the broadest sense). That this barrier may partly work can be concluded from the results of cross-breeding experiments carried out by Crapon de Caprona and Fritzsch (1984).

The same might be due in the course of differently developing feeding habits which may come into action by microseparation within the borders of large lakes with distinct bottom structures or coastlines, where rocky parts are, for instance, interrupted by sandy or muddy ones. Additionally, these differences may be combined with deep or shallow water, with or without submersed plants etc. The consequence will in any case be that non-adapted specimens will finally become the losers, and the gene flow will come to an end.

In those examples in which differently developing adaptations are coincidally combined with distinct times and spaces of reproduction—as known from *Salmo letnica* of Lake Ohrid—intralacustrine speciation seems to be the most probable manner of species formation. The results of investigations on enzyme pattern in haplochromines from Lake Victoria support this conclusion: "... a recent computer simulation study strongly suggests that disruptive selection on habitat preference, accompanied by habitat-based assortive mating, would rapidly establish reproductive isolation" (Verheyen et al. 1985, p. 476).

Turning back to the cichlids of old East African lakes three main strategies can be observed which in the end of their development may result in intralacustrine speciation;

- Homogamy between certain pairs of cichlids,
- selection of distinctly coloured males by females of the same species,
- differently developed feeding habits combined with corresponding colour varieties and characters.

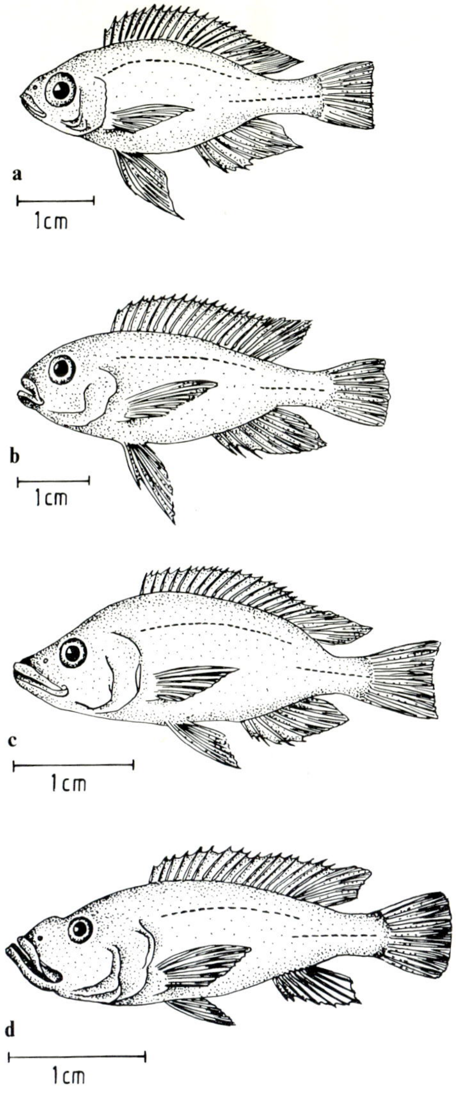

Fig. 5.2a–d. Representative *Haplochromis* ssp. from the four main feeding categories (after Greenwood 1964) (drawing M. Hänel, Hamburg). **a** *H. macrops*: generalized insectivorous species. **b** *H. sauvagei*: "winkle-picking mollusc-eater. **c** *H. parvidens*: eater of embryos and larvae, which it obtains from the mouths of brooding females. **d** *H. cavifrons*: piscivorous predator

Beginning with the last phenomenon, Greenwood has been one of the first who pointed out: "It is in their feeding habits that the *Haplochromis* have run riot, producing adaptations in dentition, skull and jaw form, and in digestive physiology. As a consequence, species are able to use almost every food source in the lake. This multifarious adaptive radiation is one of the outstanding characteristics not only of the genus *Haplochromis* but of the entire family Cichlidae" (1984, p. 260) (see main feeding types in Fig. 5.2a–d, cf. Figs. 5.3 and 5.4).

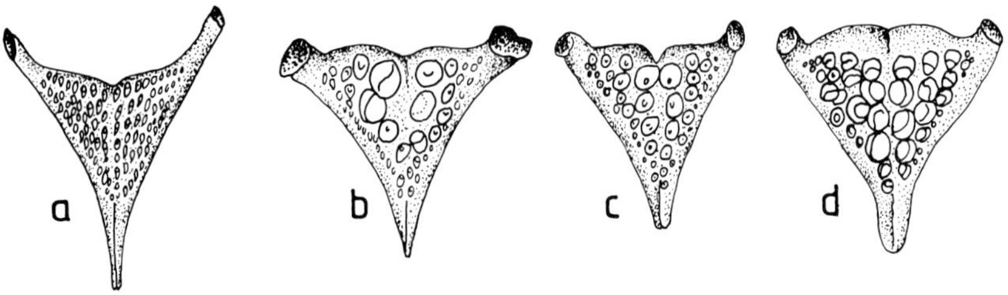

Fig. 5.3a–d. Lower pharygeal bones to show phenetic series leading from a generalized bicuspid to a specialized molariform dentition (after Greenwood 1973) (drawing M. Hänel, Hamburg). **a** *H. emodisma*: generalized type, insectivorous. **b, c** *H. humilior* and *H. obtusidens*: mixed insect and mollusc diet. **d** *H. pharyngomylus*: eats exclusively molluscs (crushing) species

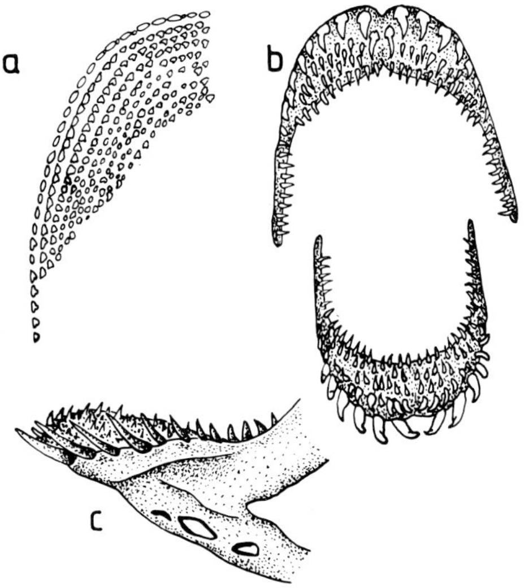

Fig. 5.4a–c. Tooth arrangement in jaws (only right premaxilla shown) (after Greenwood 1973) (drawing M. Hänel, Hamburg). **a** *H. wellcommei*: scale scraper. **b** *H. prodomus*: species that removes snails from their shells. **c** *H. xenognathus*: more specialized snail sheller

Later, other authors confirmed Greenwood's observations (Axelrod 1976; Fernald and Hirata 1977; Nelissen 1977; Witte and Witte-Maas 1981; Van Oijen 1982; Hoogerhoud et al. 1983; Brichard 1992). This kind of adaptation can be accepted without contradiction as an existing type of segregation which may result in speciation, especially when associated with ecological isolation.

The second type of stepwise segregation may have been introduced by selective breeding between *Tropheus* ssp. females and certain colour varieties of "their males". If this preference of certain colour varieties could be proved true, the assumption could be made that the obeserved selective breeding might be another very acceptable mode of speciation. It works more successfully if the

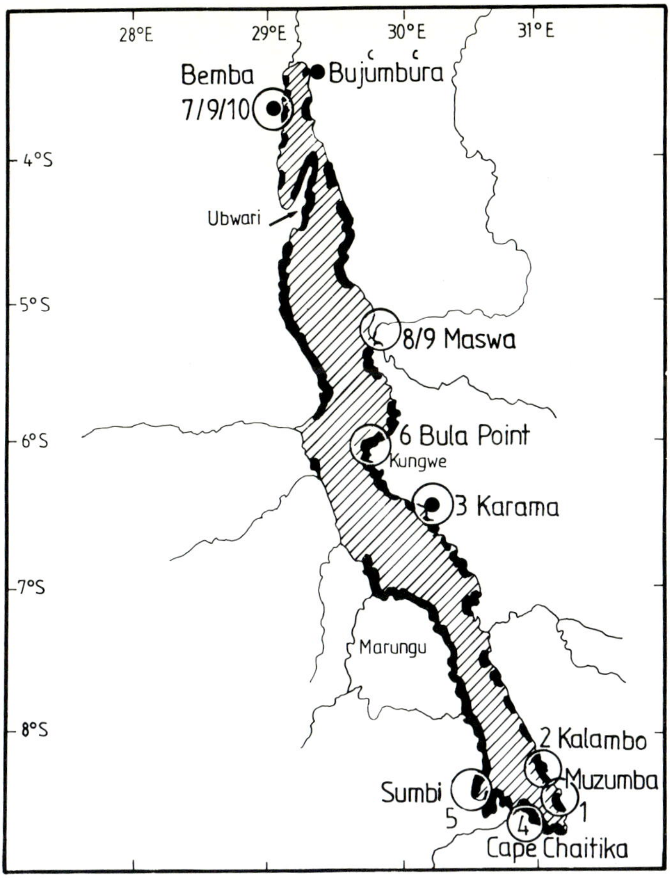

Fig. 5.5. Map of Lake Tanganyika (combined from Herrmann 1990; Brichard 1992) —— Rocky habitats; —— ecological barriers: swamps and sandy beaches

different male colour morphs are segregated to distinct parts of the lake, as is obviously the case in some *Tropheus* ssp., ranging in separate areas of Lake Tanganyika (Nelissen 1977, Brichard 1992 (Fig. 5.5; Plate I, 1–2 and 4–7; Plate II, 8–10). If additionally the female offspring follows the maternal mode of preferring colour morphs by imprinting or heredity, the success could be perfect. However, no controlled investigations have been carried out in sufficient numbers to solve these more or less open questions.

Even in cases of similar male colour morphs, possibilities of population segregation as initiation for further ongoing speciation may arise. In this connection, the very special male markers may be recalled, by which the females are forced to pair without error with only the "correct" male, e.g. the egg model on the male's anal fin (Fig. 5.6a–e). The sense of these egg models is the following: the egg-laying female tries to pick up the egg model, "thinking" it might be one

1, 5

2, 6

4, 7

Plate I. Distribution of some *Tropheus moori* varieties (see map, Fig. 5.5). *T. moori* "formenkreis": **1** *T. moori moori* (Muzumba variety); **2** *T. moori* (Kalambo variety); *T. moori* (without picture); **4** *T. moori kasabae* (Cape Chaitika variety); **5** *T. moori* (Sumbi variety); **6** *T. moori* (Bula Point variety); **7** *T. moori* (Bemba variety)

of its own eggs. By picking at the male's anal fin, the male immediatedly starts to segregate its sperm. In this complicated way, the eggs already in the mouth of the female become fertilized. These very specific behavioural patterns act as well as different colour morphs in a population or in species segregating sense (cf. Brichard 1992).

The last segregating mechanism is the so-called homogamy in cichlids, where females and males are mostly indistinguishable from each other. Nobody really knows the selective agent that attracts a certain male and a certain female selectively from alike-looking specimens. However, the fact as such is doubtless

8

9

10

Plate II. Distribution of some representatives of the *T. duboisi* "formenkreis" (see map, Fig. 5.5). **8** *Tropheus duboisi* (Maswa variety); **9** *T. duboisi* (Bemba variety); **10** *T. duboisi*: same juvenile type to nos. **8** and **9**. (photos by H.-J. Herrmann, Hamburg)

correct. Forming a pair, the latter will remain together for many spawning periods or perhaps "for ever". Evidently, this procedure will form minor populations by which gene flow within a given species is very limited and finally may lead to quick step speciation.

Already in 1947, Kosswig tried to give a genetic interpretation concerning this very special behavioural pattern and its consequences for speciation. After having taken new results into consideration he postulated: "The material basis for this type of homogamy is unknown. Nevertheless, if homogamy should be a reality under natural conditions, an understanding of the fast evolution of the cichlids in the great East African lakes would be facilitated. The splitting

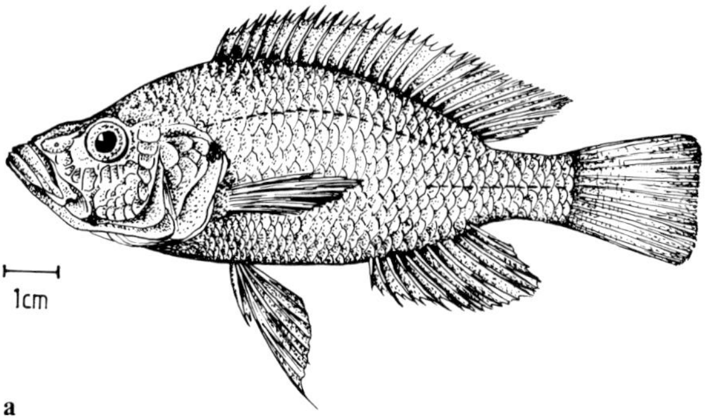

a

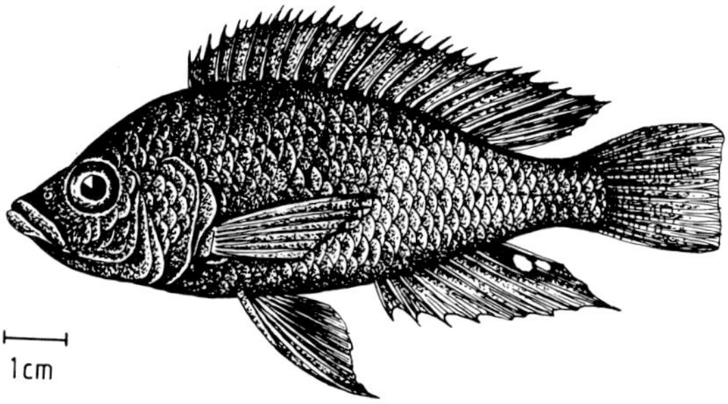

b

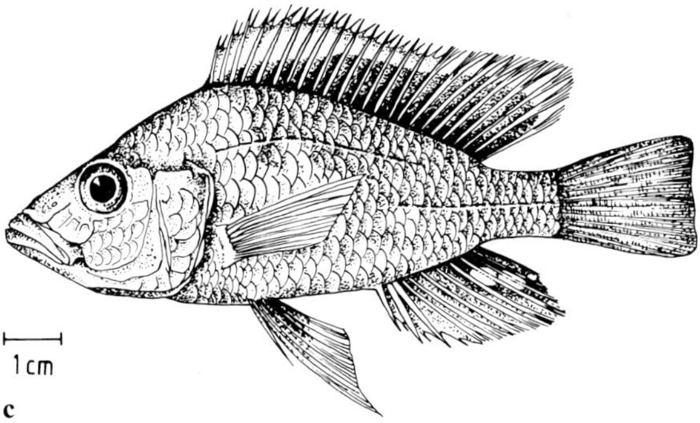

c

up of each species into some inbreeding families, each family bearing special genes for homogamic pairing, will result in the formation of separate units, morphologically alike for the systematist, but different from one another in their mating types. Each "species" of cichlids would not represent a panmictic community, but merely a number of genetically determined smaller real species" (1963a, p. 242).

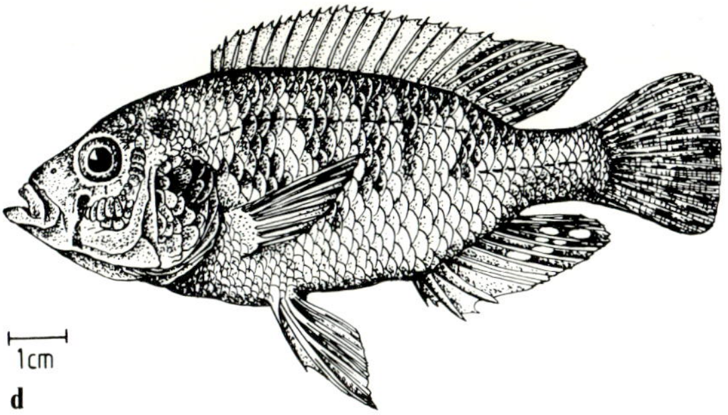

d

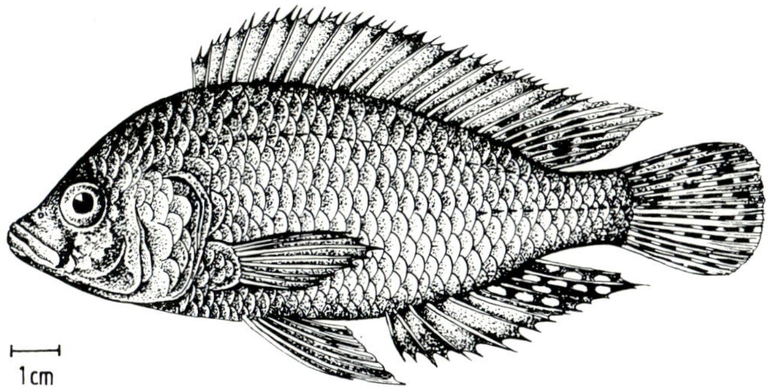

e

Fig. 5.6a–e. Anal fin with "egg model" presenting haplochromine species (after Greenwood 1965) (drawing M. Hänel, Hamburg). **a** *Hplochromis venator*: "zero" type. **b** *H. nubilus*. **c** *H. simpsoni*. **d** *H. beadlei*. **e** *Astatoreochromis alluaudi*

These ideas have not been proved wrong up to the present, even in the light of the most recent and still continuing investigations of these highly surprising form of "explosive speciation". Unfortunately, completely convincing observations on individually marked females and males have not yet been carried out, either in nature or in aquaria, the results of which might confirm or refuse the genetic interpretation first proposed by Kosswig.

In the author's personal opinion, many aspects of Kosswig's hypothesis seem to be reliable in the sense of "highly probable", but it remains an open question whether it reflects the whole truth. According to Greenwood and others, separation of lagoons and shallow parts in Lake Victoria may have caused more species segregation or at least acceleration than all the different intralacustrine abilities together. However, the theory of geographic separation covers all the observed phenomena of segregation as little as the intralcustrine explanation. Especially in large and deep lakes, e.g. Lake Tanganyika and Lake Malawi, ecological and ethological differences should be mainly responsible for the different stages of species segregation observed today, dating back to about 200 000 years ago. In this connection the following seems worthy of citation.

"The numerous endemic species and "colour forms" of Mbuna on the recently inundated rocky habitats of southern Lake Malawi confirm the long-held view that speciation in separate water bodies has been of no importance in Lake Malawi in comparison with intralacustrine speciation" (Owen et al. 1990, p. 545).

5.4 Conclusion

The stressed opinion of polyphyletic origin of the majority of the different African cichlid species groups—due to allopatric species formation—seems to be proved false for Lake Victoria, including the Lake Malawi cichlids: Meyer et al. (1990) suggest from the results of mitochondrial DNA sequences that the Lake Victoria cichlids are of monophyletic origin. According to these authors, they have derived from one species flock which segregated by selective breeding, i.e. by intralacustrine speciation processes.

References

Aksiray F (1948) Türkische Cyprinodontiden. I. Rev Fac Sci Univ Istanbul Ser B 13: 97–142

Aksiray F, Villwock W (1962) Populationsdynamische Betrachtungen an Zahnkarpfen des südwestanatolischen Aci- (Tuz-) Gölü. Zool Anz 168(1–4): 87–101

Axelrod HR (1976) Fishery resources of Lake Malawi. Presented at Congrès Européen des Ichthyologistes, Paris, 8 Sept 1976

Brichard P (1992) Tanganjika Cichliden. Bede Verlag, Kollnburg, 543 pp

Corcillius B (1979) Weitere Untersuchungen über Speziationsphänomene in der Neuweltlichen Zahnkarpfengattung *Cyprinodon* (Pisces, Cyprinodontidae). Diss, Universität Hamburg, 96 pp

Crapon de Caprona MD, Fritzsch B (1984) Interspecific fertile hybrids of haplochromine Cichlidae (Teleostei) and their possible importance for speciation. 34(4) 503–538

Fernald R, Hirata N (1977) Field study of *Haplochromis burtoni*: habitats and co-habitant. Environ Biol Fishes 2(3): 299–308

Fryer G (1959) The trophic interrelationships and ecology of some littoral communities of Lake Nyasa with especial reference to the fishes, and a discussion of the evolution of a group of rock-frequenting Cichlidae. Proc Zool Soc Lond 32: 153–281

Greenwood PH (1964) Explosive speciation in African lakes. Proc R Inst 40(184): 256–269

Greenwood PH (1965) The cichlid fishes of Lake Nabugabo, Uganda. Bull Br Mus (Nat Hist) Zool 12(9): 315–357

Greenwood PH (1973) Morphology, endemism and speciation in African cichlid fishes. Verh Dtsch Zool Ges 66: 115–124

Herrmann HJ (1990) Die Buntbarsche der Alten Welt. Tanganjika-See. Ulmer Stüttgart, 224 pp

Hoogerhoud RJC, Witte F, Barel CDN (1983) The ecological differentiation of two closely resembling *Haplochromis* species from Lake Victoria (*H. iris* and *H. hiatus*; Pisces, Cichlidae). Neth J Zool 33(3): 283–305

Kohzov M (1963) Lake Baikal and its life. Junk, The Hague, 344 pp

Kosswig C (1947) Selective mating as a factor for speciation in cichlid fish of East African lakes. Nature 159: 604–605

Kosswig C (1956) Über Makro- und Mikropopulationen des Zahnkarpfens *Anatolichthys*. Zool Anz 156(5/6): 75–90

Kosswig C (1963a) Ways of speciation in fishes. Copeia 2: 238–244

Kosswig C (1963b) Evolutionsphänomene bei Knochenfischen, genetisch betrachtet. Veröff Inst Meeresforsch Bremerhaven Drittes Meeresbiol Symp: 178–196

Kosswig C, Sözer F (1945) Nouveaux Cyprinodontides de l'Anatolie centrale. Rev Fac Sci Univ Istanbul Ser B 10: 77–83

McElroy DM, Kornfield I (1990) Sexual selection, reproductive behavior, and speciation in the mbuna species flock of Lake Malawi (Pisces: Cichlidae). Environ. Biol Fishes 28: 273–284

Meyer A, Kocher THD, Basasibwaki P, Wilson AC (1990) Monophyletic origin of Lake Victoria cichlid fishes suggested by mitochondrial DNA sequences. Nature 347: 550–553

Nelissen MHJ (1977) Contribution to the ethology of *Simochromis babaulti* Pellegrin (Pisces, Cichlidae). Ann Soc Zool Belg 106(2–4): 167–175

Owen RB, Crossley R, Johnson TC, Tweddle D, Kornfield I, Davison S, Eccles DH, Engstrom DE (1990) Major low levels of Lake Malawi and their implications for speciation rates in cichlid fishes. Proc R Soc Lond B240: 519–553

Rosenthal H (1993) Annotated bibliography on transplantations and transfers of aquatic organisms and their implications on aquaculture and ecosystems. (in press)

Stankovic S (1953) Sur Ia speciation dans le lac d'Ohrid. Verh Int Ver Limnol (Lond) 12: 478–506

Trewavas E (1947) Speciation in cichlid fishes of East African lakes. Nature 160: 96–97

Trewavas E (1948) The origin and evolution of the cichlid fishes of the great African lakes, with special reference to Nyasa. Rapp. 13th Congr Int Zool (Paris) Sect 5b: 1–4

Van Oijen, MJP (1982) Ecological differentiation among the piscivorous haplochromine cichlids of Lake Victoria (East Africa) Neth J Zool 32(3): 336–363

Verheyen E, Rompaey J Van, Selens M (1985) Enzyme variation in haplochromine cichlid fishes from Lake Victoria. Neth J Zool 35(3): 469–478

Villwock W (1958) Weitere genetische Untersuchungen zur Frage der Verwandtschaftsbezeihungen anatolischer Zahnkarpfen. Mitt Hamb Zool Mus Inst 56: 81–152

Villwock W (1963) Genetische Analyse des Merkmals "Beschuppung" bei anatolischen Zahnkarpfen (Pisces, Cyprinodontidae) in Auflöserversuch. Zool Anz 170(1/2): 23–45

Villwock W (1986) Speciation and adaptive radiation in Andean *Orestias* fishes. In: Vuilleumier F, Monasterio M (eds) High altitude tropical biogeography. Oxford University Press, New York, pp 387–403

Villwock W (1993) Die Titicaca-See-Region auf dem Altiplano von Peru und Bolivien und die Folgen eingeführter Fische auf Wildarten und ihren Lebensraum. Naturwissenschaften 80: 1–8

Witte F, Witte-Maas ELM (1981) Haplochromine cleaner fishes: a taxonomic and ecomorphological description of two new species. Revision of the haplochromine species (Teleostei, Cichlidae) from Lake Victoria. Part I. Neth J Zool 31(1): 203–231

6 Modelling Extinction and Survival of Small Populations

C. WISSEL, T. STEPHAN, and S.-H. ZASCHKE

6.1 Introduction

When one speaks of small populations, the problem of their survival is the first point of interest. It is well known (Shaffer 1981; Soulé 1987) that small populations are more likely to die out. One may be interested in the size of a habitat which is sufficient to provide a good chance of survival; but limitations of resources other than space can be of decisive influence. We do not speak of deterministic extinction, when the mean environmental conditions are insufficient for a longer survival of a population. Instead, we consider situations which provide good conditions of existence on average; but the temporal alterations and random events can put populations at risk of extinction, especially small populations, as will be seen below.

Man has fragmented the landscape and has shrunken the habitats of population also in other ways. From the point of view of conservation, one is interested in minimal viable populations (MVP) (Shaffer 1981). Quantitative statements are necessary in nature conservation. For this purpose, mathematical models are suitable. Based on demographic data, they may help to make prognoses. They can be used to obtain fundamental insight into long-term changes of survival. However, one should not expect recipes for biological conservation; the problems of extinction are too complex, and models can never include all influences (Wissel 1992). They can help to identify decisive risk factors and to find general trends.

This chapter shows the contribution of theoretical ecology (mathematical models) in investigating the chance of survival of small populations. The influence of the carrying capacity K, which is often related to the area available to the population, is of great interest (Shaffer 1981; Soulé 1987), but other factors also have decisive influence on the chance of survival. We will show that only probabilistic statements (Soulé 1987; Wissel and Stöcker 1991) are possible because random influences, such as unpredictable fluctuations of the environment, variability of the individuals and rare catastrophic events, play the decisive role. This means that extinction can never be excluded with total certainty; there is always a possibility of risk. Science cannot decide which risk is tolerable; this is a question of politics. Given a tolerable risk of extinction for a certain period of time, the corresponding minimal viable population (MVP) (Shaffer 1981) may be determined, i.e. the size of a population whose probability of extinction is not greater than the predetermined threshold (risk). Or, given the size of a population, what is the probability that it becomes extinct in a certain period of time? This chapter will show which populations are especially vulnerable.

6.2 Population Dynamics with Overlapping Generations

6.2.1 Environmental and Demographic Noise

The usual description (Wissel 1989a, b; Wissel and Stöcker 1991) of population dynamics continuous in time is

$$dN/dT = Nr(N). \tag{1}$$

This description applies for populations with overlapping generations. In this case, birth and death processes are always possible. The change dN of the individual number N per time interval dt is determined by the growth rate on the right side of Eq. (1). Therefore, r(N) is the individual growth rate, i.e. r(N) is the difference of the birth rate and the death rate of one individual, which can be determined from demographic data.

It is well known that there must be a density-dependent regulation at high individual numbers N. The simplest mathematical possibility of a density-dependent regulation is the logistic growth which is given by the first term of

$$r(N) = r(1 - N/K) + \sigma f(t). \tag{2}$$

As mentioned above, the temporal variation of the environment is important. This means that birth and death rates change in the course of time because of varying environmental conditions. Thus we add (Goel and Richter-Dyn 1974; Wissel 1989a; Wissel and Stöcker 1991) a temporal varying part $\sigma f(t)$ to the mean growth rate $r(1 - N/K)$. The amplitudes of this variation are given by σ. Consequently, f(t) describes the temporal fluctuation of the growth rate with unit amplitudes.

There are an immense number of temporally changing factors which may influence the dynamics of a population. Neither are all of them known, nor can we include them all explicitly in the model. The only possibility is to subsume their influence under the fluctuating contribution $\sigma f(t)$ to the growth rate. As the temporal variation of these factors is unknown, we can describe them only as random influences. Therefore $\sigma f(t)$ is called environmental random noise or environmental stochasticity. In general, it is assumed (Goel and Richter-Dyn 1974; Wissel 1989a) that f(t) is normally distributed and that there is no correlation between the values of f(t) at different instants of time. The last assumption is valid if the time scale of the environmental fluctuations is shorter than the typical time scale of the population dynamics, i.e. shorter than 1/r (Wissel 1989a; Wissel and Stöcker 1991).

Equations (1) and (2) are a reasonable description of population dynamics with environmental noise at high individual numbers N (Wissel and Stöcker 1991). At low individual numbers the description of the integer individual number by a continuous variable N is not appropriate. This means that one can find only numbers like 0, 1, 2, 3 and not 1.37. Therefore, Eq. (1) cannot be used at small values of N. But this is just the range of N which is of interest for small populations and extinction problems,· therefore another description is necessary for small numbers N.

A suitable probabilistic description of population dynamics, which uses only integer values of the individual number, are the so-called stochastic birth and death process (Goel and Richter-Dyn 1974; Soulé 1987; Wissel and Stöcker 1991). Let $\lambda_n \cdot \Delta t$ be the probability that one of n present individuals gives birth in a short time interval Δt, and $\mu_n \Delta t$ be the probability that one of n present individuals dies in Δt. Given these birth and death rates, λ_n and μ_n, respectively, one can determine the probability $P_n(t)$ of having n individuals at the time t from the following equation [Eq. (3)] in principle (Goel and Richter-Dyn 1974).

$$dP_n/dt = \lambda_{n-1} P_{n-1} + \mu_{n+1} P_{n+1} - (\lambda_n + \mu_n) P_n. \qquad (3)$$

The dependence of the birth and death rates, λ_n and μ_n, on the present individual number n is decisive for this stochastic process.

In Fig. 6.1 an example is shown with birth and death rates given by Eqs. (8) and (9). Instead of calculating a probability, a so-called Monte Carlo simulation is carried out. This means that the individual number is increased or decreased in the course of time using a random generator which determines the appearance of birth and death in correspondence with the probabilities λ_n and μ_n. Random fluctuations of the individual numbers are seen, which are said to result from demographic noise or demographic stochasticity (Goodman 1987b; Wissel and Schmitt 1987; Wissel and Stöcker 1991). The reason for this is the variability of the individuals; they are not little machines which give birth and death in a regular and predictive manner. Only probabilities for birth and death can be given because the detailed conditions for a single birth or death event are not known. Therefore the unpredictable order of birth and death events gives rise to the demographic fluctuations seen in Fig. 6.1.

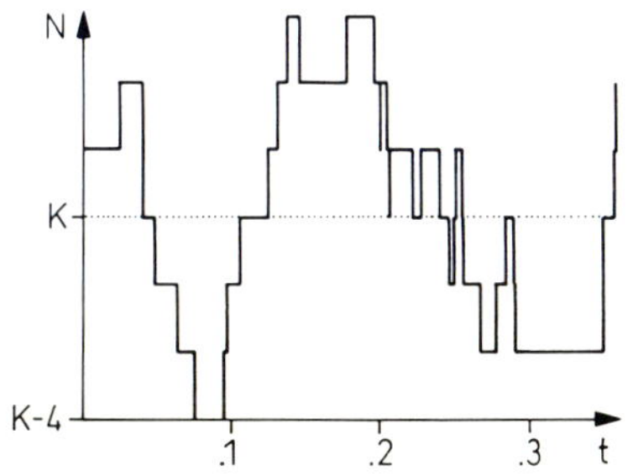

Fig. 6.1. Monte Carlo simulation of fluctuating population dynamics with demographic noise. Individual number N versus time t for birth and death rates given by Eqs. (8) and (9) with $K = 50$ and $R = 2r$

At small individual numbers, one can argue in the following way (Wissel and Schmitt 1987; Wissel and Stöcker 1991): if the individual number n is far below the carrying capacity of a habitat, the limitation of resources plays no role and there is no intraspecific competition. Therefore no density dependence is expected for the birth and death rates. This means that the birth rate λ_1 and death rate μ_1 per individual should be independent of the number n of conspecifics. Thus, the birth rate λ_n and death rate μ_n of the whole population are

$$\lambda_n = n\lambda_1 \tag{4}$$

$$\mu_n = n\mu_1. \tag{5}$$

The difference

$$\lambda_n - \mu_n = n(\lambda_1 - \mu_1) = nr \tag{6}$$

gives the growth rate nr [see Eq. (1)] for small individual numbers. Thus $\lambda_1 - \mu_1$ equals the intrinsic growth rate r [see Eq. (2)]. The sum

$$\lambda_n + \mu_n = n(\lambda_1 + \mu_1) = nR \tag{7}$$

is known (Wissel and Schmidt 1987; Goodman 1987b; Wissel and Stöcker 1991) to give the variance of the growth, i.e. it determines the amplitudes of the random fluctuations seen in Fig. 6.1. As it results from the variability of the individuals, it is called demographic noise. The abbreviation R was introduced for the sum $\lambda_1 + \mu_1$, which describes the strength of the demographic noise. It should be emphasized that this demographic noise arises without any environmental variations in the model, only by the variability of the individuals.

The density independence of the individual birth rate λ_1 and death rate μ_1 can only be valid at small individuals numbers. It is a logical need to have a density-dependent regulation at higher values of n, otherwise an unlimited growth would result. The simplest possibility of a density dependence is the logistic form [see Eq. (2)] for the individual growth rate r(N). This means

$$\lambda_n = \tfrac{1}{2}[rn(1 - n/K) + Rn] \tag{8}$$

$$\mu_n = \tfrac{1}{2}[- rn(1 - n/K) + Rn]. \tag{9}$$

These definitions give the logistic form for the difference $\lambda_n - \mu_n$ as it should be for the growth rate Nr(N) which equals this difference. The sum is unchanged, giving the demographic noise Rn.

A model which combines both demographic [Eq. (7)] and environmental [Eq. (2)] noise can be found in two different ways (Wissel and Schmitt 1987; Goodman 1987a, b; Wissel and Stöcker 1991). It reads

$$\lambda_n = \tfrac{1}{2}[rn(1 - n/K) + Rn + \sigma^2 n^2] \tag{10}$$

$$\mu_n = \tfrac{1}{2}[- rn(1 - n/K) + Rn + \sigma^2 n^2]. \tag{11}$$

The difference gives the logistic growth as it should be. In the sum which describes the noise, both demographic noise Rn and environmental noise $\sigma^2 n^2$

appear. The strength σ of the environmental noise was defined in Eq. (2). Equations (10) and (11) can be deduced from Eqs. (1) and (2) (Wissel and Schmitt 1987; Wissel and Stöcker 1991) if the correlation time of the environmental fluctuations is smaller than the time scale 1/r of the population dynamics and smaller than a proper time step of these dynamics. In this case the so-called Ito calculus has to be applied (Goel and Richter-Dyn 1974; Turelli 1977). The proper time step of population dynamics may be the time an individual needs to grow from birth to an adult, fertile individual. If this correlation time is smaller than 1/r but greater than the proper time step, the so-called Stratonovich calculus has to be chosen. In this case (Goel and Richter-Dyn 1974), the intrinsic growth rate r has to be substituted by

$$r \rightarrow r + \tfrac{1}{2}\sigma^2. \tag{12}$$

This has to be used in Eqs. (10) and (11). A simpler way (Goel and Richter-Dyn 1974) of limiting growth of a population is to assume density-independent birth and death rates until the carrying capacity K is reached. There the growth is stopped by setting $\lambda_k = 0$. In this case we obtain, instead of Eqs. (10) and (11),

$$\lambda_n = \tfrac{1}{2}[rn + Rn + \sigma^2 n^2] \quad \text{for } n < K \tag{10a}$$

$$\mu_n = \tfrac{1}{2}[-rn + Rn + \sigma^2 n^2]. \tag{11a}$$

This form of limited growth will later be used to investigate the influence of different types of density regulation.

Now the model Eq. (3) with Eqs. (10) and (11) for population dynamics with environmental and demographic noise is complete (Wissel and Schmitt 1987; Goodman 1987a, b). Before using it for calculations, the parameter values have to be assessed. We start from the individual birth rate λ_1 and death rate μ_1 without environmental noise at small individual numbers where no density dependence appears. Appendix A shows how they can be determined from demographic data. The difference $\lambda_1 - \mu_1$ gives (Goodman 1987a, b; Wissel and Schmitt 1987; Wissel and Stöcker 1991) the intrinsic growth rate r [see Eq. (6)]. The sum $\lambda_1 + \mu_1$ gives the strength R of the demographic noise [see Eq. (7)]. The carrying capacity K should be proportional to the amount of available limiting resources. This may be space, i.e. area or volume, or other resources like food or water. One can often argue that the extent of resources is proportional to the area or volume of a habitat.

The assessment of the strength σ of the environmental noise is more difficult. Appendix B shows how the mean value m(ΔN) of the increase ΔN of the individual number N and its standard deviation Sd(ΔN) can be approximated (cf. Goodman 1987b) for a short time interval, Δt, at high individual numbers N. If only the environmental noise and density-independent growth are considered then

$$\frac{Sd(\Delta n)}{m(\Delta n)} = \frac{\sigma}{r}\frac{1}{\sqrt{\Delta t}} \tag{13}$$

Therefore, if one can estimate this variation in growth for a time interval Δt and if the intrinsic growth rate, r, is known, one obtains an estimate for the strength of the environmental noise. The longer the time interval the more the variation [Eq. (13)] is smoothed out.

Let us consider a population with a growth rate of $r = 0.1$/year. For this slowly growing population, one may consider a time interval of $\Delta t = 10$ years. Let the relative standard deviation for 10 years be $Sd/m = 0.2$. From Eq. (13) we find $\sigma^2/r = 0.04$. On the other side consider $r = 5$/year. This gives an increase in the individual number during 1 year by a factor $e^5 \approx 150$ because of the exponential growth. Let us consider $\Delta t = 1$ year and a relative standard deviation for 1 year of $sd/m = 0.8$, then we find $\sigma^2/r = 3.2$. These values should be the extreme limits for the range of the strength σ of the environmental noise.

The relative importance of demographic noise Rn and environmental noise $\sigma^2 n^2$ can be assessed by considering the quotient

$$\frac{\sigma^2 n^2}{Rn} = \frac{\sigma^2 n}{R}. \tag{14}$$

We see that for a sufficiently high individual number n, the environmental noise is decisive and demographic noise plays no role. For small individual number n, the size of σ^2 compared to R is essential. If σ^2 is considerably smaller than R, only demographic noise is important for small n, and environmental noise is irrelevant.

6.2.2 Results

6.2.2.1 Quantities of Interest

Next we introduce the quantities which can be calculated from our model in Eqs. (3), (10) and (11), and which are of interest for survival or extinction of small populations. The mean life time T_n (time until extinction) can be calculated by the formula (C1) in Appendix C. It is the mean survival time of a population which contains n individuals at the moment. The mean life time T_K of a population which is established at the carrying capacity K is of special interest. K is the deterministic equilibrium value where the individual number N is expected to be found most of time, at least for small fluctuations (see Fig. 6.2).

It can be shown (see Appendix D) that the probability $P_n(t)$ of having n individuals at time t goes over to a quasistationary distribution P_n^* in the course of time. There the relative frequency of different individual numbers does not change, presupposing that the population has not been extinguished so far. An example is shown in Fig. 6.2. The distribution P_n^* describes the probability of different amplitudes of the fluctuations seen in Fig. 6.1. One can look upon P_n^* as the frequency distribution of finding distinct individual numbers n in the course of time, following a time series like Fig. 6.1.

An important problem is the assessment of the chance that a population threatened by the reduction of its individual numbers may recover. This can

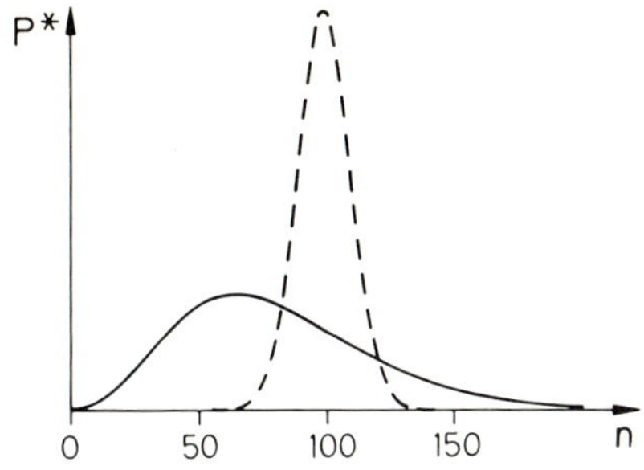

Fig. 6.2. Quasi-stationary probability distribution P^* of having n individuals in a population described by Eqs. (10) and (11) with $K = 100$ and $R = 2r$. *Dashed line* without environmental noise ($\sigma = 0$) and *full line* for environmental noise of $\sigma^2 = r/3$

be described by the probability R_n that a population which contains n individuals at the moment will recover to an established state, i.e. that it returns to the carrying capacity K. The formula for R_n is given in Appendix C [Eq. (C3)]. R_n can be looked upon in two different ways; one may consider a population reduced to the individual number n by a disturbance. Then R_n describes the probability of recovery. Or you may consider a population which has reached the individual number n from a colonizing state. Then R_n describes the probability that this colonizing population will reach the established state at $n = K$. For a colonizing individual, one may be interested in the probability W_n for reaching the individual number n before becoming extinct. This quantity describes the success of a colonizing individual. The formula for this quantity W_n is given by Eq. (C3) in Appendix C.

Finally, we consider the probability $P_0(t)$ that the individual number is zero, i.e. that the population is already extinct at the time t. For the determination of this quantity the complete time-dependent solution of Eq. (3) with Eqs. (10) and (11) is necessary. A numerical method (Wissel 1984, 1989a) is described in Appendix D. This probability $P_0(t)$ is the central quantity of interest. As the time of final extinction of a population is not exactly known, only probabilistic considerations are possible. For very short periods of time one expects that the probability of extinction is very small. However, in the long term, the probability of extinction increases. The concept (Shaffer 1981) of minimum viable population (MVP) can only be defined in a reasonable manner by use of this probability $P_0(t)$. As there is a remaining risk of extinction for all instants of time, one can only ask for a size of a population which guarantees that the risk (probability) of extinction does not exceed a given limit; but as this probability increases in the course of time, one has to fix an instant of time where one considers the risk. Thus, a reasonable definition (Shaffer 1981) of minimum viable populations (MVP) in a given environment is the population with the smallest carrying capacity K having a chance of, say, 99% of survival for 100 years. In other words, the size of the carying capacity must be sufficient to guarantee that the probability of extinction $P_0(t)$ at the time $t = 100$ years is not greater than 0.01.

$$P_0(t = 100 \text{ years}) \leqslant 0.01. \tag{15}$$

It has to be emphasized that there is no special reason to choose 100 years for the time span and a remaining risk of 1%. One can equally as well take, say, 300 years and 3%. This ambiguity in the definition of MVP is unavoidable. It is a consequence of the fact that only a probabilistic statement is possible. There is no scientific way of choosing a particular time or remaining risk. It is a question of evaluation, which risk of extinction one tolerates at which time.

6.2.2.2 Qualitative Results

First we discuss the importance of the environmental noise $\sigma^2 n^2$. For this purpose, an example of the quasistationary distribution P_n^* is shown in Fig. 6.2. Without environmental noise we find a peaked distribution around the carrying capacity $K = 100$. This indicates that the fluctuating individual number n remains in the neighbourhood of K and that there is almost no chance of finding individual numbers near zero. A moderate environmental noise changes the situation completely. Now there is a chance of finding very small individual numbers. The importance of environmental noise becomes more obvious by looking at the mean life time T_K of a population which is established at the carrying capacity at the moment. For the parameter values of Fig. 6.2 we obtain $T_K = 5.9 \times 10^{21}/r$ in the case without environmental noise and $T_K = 9.9 \times 10^3/r$ for the case of $\sigma^2 = r/3$. As biologically realistic values of r lie in the range from $r = 0.1$/year to $r = 10$/year, we see that without environmental noise there is almost no risk of extinction for this population in an ecologically interesting time span, whereas the mean survival time T_K may be rather short with environmental noise.

Besides the environmental noise, the size of the carrying capacity K plays a decisive role. In Appendix C it is shown that the mean life time T_K of an established population ($n = K$) grows with the carrying capacity K asymptotically as

$$T_K = \text{const}\, K^{\alpha - 1} \quad \text{for } \alpha > 1 \tag{16}$$

$$T_K = \text{const} \ln K \quad \text{for } \alpha < 1 \tag{17}$$

with

$$\alpha = 2r/\sigma^2 \quad \text{(Ito calculus)} \tag{18a}$$

$$\alpha = 2r/\sigma^2 + 1 \quad \text{(Stratonovich calculus).} \tag{18b}$$

Remember that the correlation time of the environmental noise in relation to a proper time step of the population dynamics determines if the Ito or Stratonovich calculus has to be chosen. For a longer correlation time, the Stratonovich approach is appropriate and Eqs. (18b) and (16) consequently apply. In this case, the logarithmic increase in T_K is not found. Only if the correlation time is shorter than the proper time step of the population dynamics does Ito's approach have to be taken [see Eq. (18)]. In this case a logarithmic increase in T_K with K is found only for larger environmental noise ($\sigma^2 > 2r$) (cf. Goodman 1987a, b). In order to see where this asymptotic behaviour is reached, one can compute T_K for different values of K by the help of formula

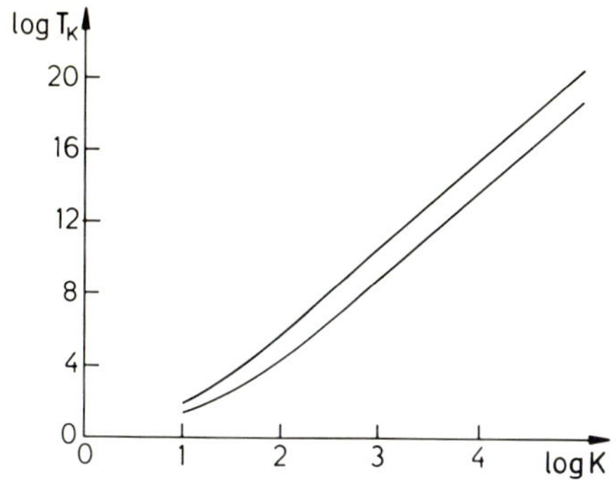

Fig. 6.3. Mean life time T_K of a population versus carrying capacity K in double logarithmic scale. *Lower curve* for model in Eqs. (10) and (11) and upper curve for model in Eqs. (10a) and (11a) with $r = 1$, $R = 2$, $\sigma^2 = 1/3$ and time in arbitrary units

(C1) in Appendix C. A typical result is shown in Fig. 6.3. There a double logarithmic scale is used which transforms Eq. (16) to

$$\log rT_K = \log \text{const}' + (\alpha - 1)\log K. \tag{16a}$$

One sees that the asymptotic behaviour is reached at about $\log K = 2$, i.e. $K = 100$. The same order of magnitude for this point where the asymptote is reached is found for other parameter values as well.

Although T_K is only a mean value (mean life time), its dependence on K gives important information for the assessment of the minimum viable population, as will be shown later. In order to obtain an impression of the effect of an increase in K, let us consider following cases. For $\alpha < 1$, we consider an increase from $K = 36$ to $K = 1296$. This gives an increase of T_K by a factor of 2 only because of Eq. (17). If an increase in the mean life time T_K by a factor 10 is wanted, an increase from K to K^{10} is needed. Other relations are shown in

Table 6.1. Relation between mean life time T and carrying capacity K

An increase from $K = 16$ to $K = 256$ gives a change in T_K by a factor f_T if $T_K = \text{const}\, K^{\alpha-1}$ is approximately valid

α	<1	1.5	2	3	5
f_T	2	4	16	256	65 536

In order to obtain an increase in T_K by a factor of 10, one needs an increase in K by a factor of f_K:

α	<1	1.5	2	3	5	9	17
f_K	K^9	100	10	3.16	1.78	1.33	1.15

The relation between α, σ^2/r and $\sigma/\sqrt{r}$ is given by

α	<1	1.5	2	3	5	9	17
σ^2/r	>2	1.33	1	0.67	0.4	0.22	0.12
$\sigma/\sqrt{r}$	>1.41	1.15	1	0.82	0.63	0.47	0.34

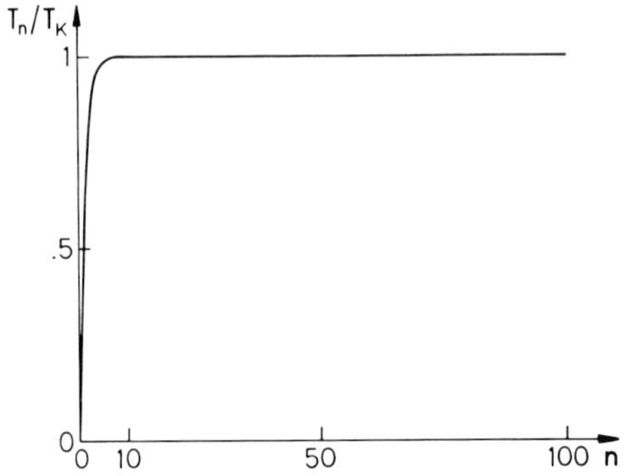

Fig. 6.4. Mean life time T_n (in units of T_K) of a population which contains n individuals at the moment versus n for model in Eqs. (10) and (11) with $r = 1$, $R = 2$, $\sigma^2 = 0.1$. $K = 100$ and time in arbitrary units

Table 6.1. Remember that in the Stratonovich approach the effective r [see Eq. (12) or Eq. (18b)] is the value of $r + 1/2\sigma^2$ and we always have $\alpha > 1$.

So far we have considered the mean life time T_K of a population which is established at the carrying capacity K. In Fig. 6.4, the mean life time T_n of a population is shown which contains n individuals at the moment. The curve shows a saturation behaviour for large n which is typical for not too strong environmental noise. This behaviour becomes more understandable when one considers Fig. 6.5, which shows the probability R_n of reaching the established state at K starting with the individual number n [see Eq. (C3) in Appendix C for calculation]. The same saturation behaviour as in Fig. 6.4 is found.

Several conclusions can be drawn for Figs. 6.4 and 6.5. There is a critical individual number n_c above which the probability of extinction is comparatively lower. To be more precise, n_c is the number at which R_n equals 0.95; this is $n_c = 6$ in Fig. 6.5. Populations with individual numbers n above n_c go to the established state at the capacity K almost in any case (Fig. 6.5). Therefore their mean life time T_n hardly differs from the mean life time T_K of an established population (Fig. 6.4). One consequence of this is that n_c gives a limit for the disturbance of an established population which can be tolerated without increased risk of extinction. This disturbance can be caused by man or may be the

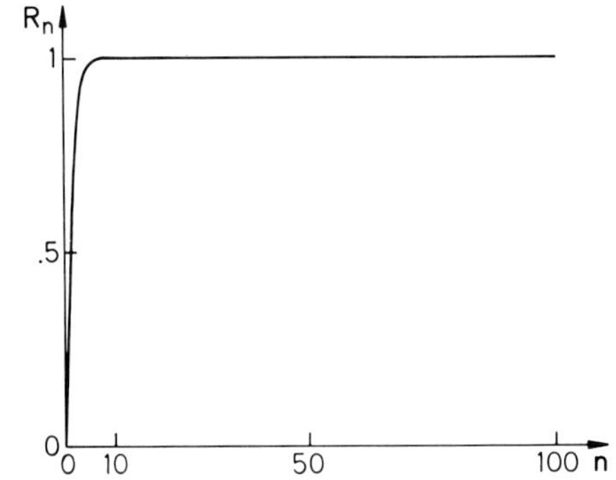

Fig. 6.5. Probability R_n of reaching the carrying capacity K if the individual number of the population is n at the moment, versus n. Model in Eqs. (10) and (11) for $r = 1$, $R = 2$, $\sigma^2 = 0.1$, $K = 100$ and time in arbitrary units

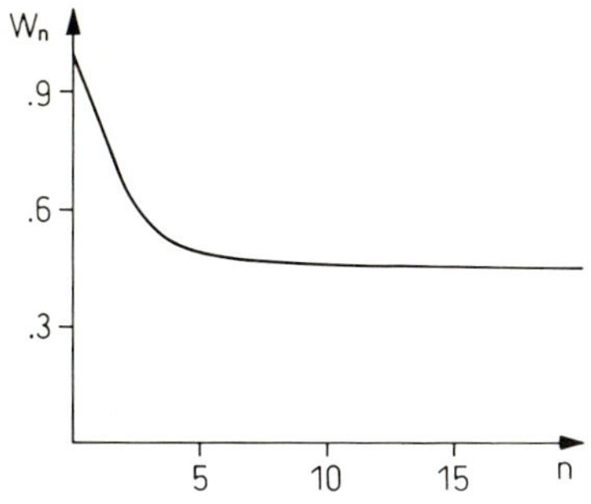

Fig. 6.6. Probability W_n that one colonizing individual will produce a population of size n versus n. Model in Eqs. (10) and (11) with $K = 20$, $R = 3r$, $\sigma^2 = 0.01r$

consequence of sporadic catastrophic natural events. The influence of the carrying capacity K on the mean life time T_K describes how large a habitat must be or how large the amount of available limiting resources (quality of a habitat) should be to provide a good chance of survival for a certain period of time. The probability R_n determines the risk of becoming extinct for a population of the actual size n.

One can, however, look upon Fig. 6.5 from a different point of view. One may consider a population in a colonizing stage consisting of n individuals. Then R_n gives the probability of the success of the colonization, i.e. the probability of reaching the established state at the carrying capacity K. Below n_c the success of colonization is unsure and is the smaller, the smaller the number n of colonizers is. Above n_c the population has survived this critical stage of colonization and will become fully established. Similar information [see Eq. (C3) in Appendix C for calculation] is obtained from the probability W_n that a single colonizator will produce a population of the individual number n in the course of time. An example is shown in Fig. 6.6. The higher the individual number n, the smaller the probability W_n that this will be reached by the colonizing population. However, as in Figs. 6.4 and 6.5. a saturation is found above the critical individual number n_c larger populations than n_c will be reached with the same probability. The saturation level W_K equals R_1, i.e. it describes the probability that one colonizing individual gives rise to an established population.

The so-called additivity of T_n provides further insight into extinction processes. The difference of the mean life times $T_n - T_{n'}$ for populations with different actual individual numbers $n > n'$ is the mean time a population needs for going from individual number n down to individual number n'. Looking at Fig. 6.4, one can conclude that the fluctuations of the individual number down to smaller values are comparatively fast as long as the individual number n does not fall below n_c. Below n_c, where the risk of extinction is increased, the time to reach this lower number n is much longer.

For the assessment of minimum viable populations (MVP) the probability $P_0(t)$ of having no individual (extinction) at time t has to be used. In Fig. 6.7 $P_0(t)$ is shown for the same parameter values as in Figs. 6.4 and 6.5. At $t = T_K$

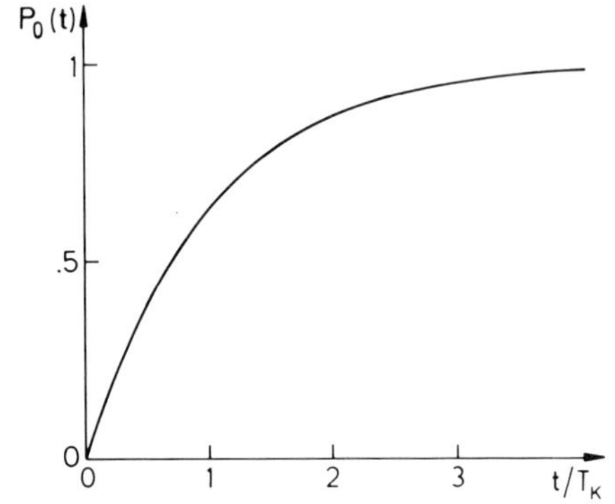

Fig. 6.7. Probability $P_0(t)$ of having no individual in the population at time t versus t in units of the mean life time T_k for model in Eqs. (10) and (11) with $R = 2r$, $\sigma^2 = 0.1r$ and $K = 100$

(mean life time), the probability $P_0(t)$ of extinction is 63%; but obviously population can die out much earlier or much later. There is a huge variability in the life time t. For a time $t = 0.05\ T_K$, the probability of extinction reaches the 5% mark. One way of using $P_0(t)$ is the consideration of a population in a given habitat which determines the value of the carrying capacity. The curve $P_0(t)$ can be used to determine the time t for which the risk of extinction remains below a given limit, say 5%. Or, given a certain period of time t, the curve $P_0(t)$ can be used to determine the risk that the population will not survive this period. For a MVP the tolerable risk and the period of time which is of interest have to be given first. Then one has to search for the smallest carrying capacity K, which guarantees that the risk does not exceed the given level in the prescribed time. Therefore, curves of $P_0(t)$ for different values of K have to be compared until the corresponding risk coincides with the given limit at the prescribed time. In order to shorten this tedious procedure, the following can be used. We considered a variety of $P_0(t)$ curves for different parameter values. $P_0(t)$ typically reaches a value between 60 and 65% at the time $t = T_K$. A value of 5% is typically obtained for times t between $0.06\ T_K$ and $0.2\ T_K$. As mathematical models can provide only rough estimations, it is sufficient to say that the 5% limit lies at $0.1\ T_K$. The dependence of T_K on the carrying capacity shows the general trend given by Eqs. (16) and (17). An example for a detailed curve is shown in Fig. 6.3, which is comparatively easy to calculate. If one wants to know the MVP for 5% risk and for a period of 100 years, one has to go to the point where $T_K = 1000$ years ($\ln T_K = 3$), and obtains (lower curve) $\log K = 1.65$, i.e. $K = 45$. This is an estimate for the smallest K (habitat), which guarantees that the risk of extinction does not exceed 5% during 100 years for a population with logistic growth (lower curve) under conditions which are characterized by the parameter values.

Finally, we consider cases with strong environmental noise. In these cases, the form of T_n and R_n differs from Figs. 6.4 and 6.5. Examples for T_n and R_n for larger strength σ of environmental noise are shown in Figs. 6.8 and 6.9. Obviously, no critical individual number n_c appears above which the population is in a relatively safe situation. The conclusion for smaller environmental noise

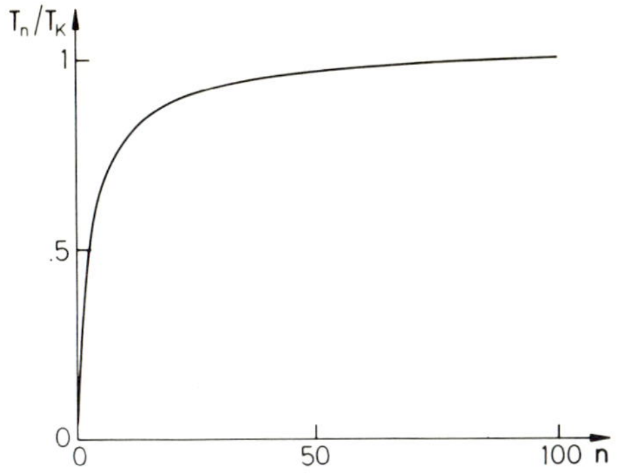

Fig. 6.8. Mean life time T_n (in units of 1/r) of a population of the size (individual number) n versus n for model in Eqs (10) and (11) with $R = 2r$, $\sigma^2 = 1.5r$ and $K = 100$

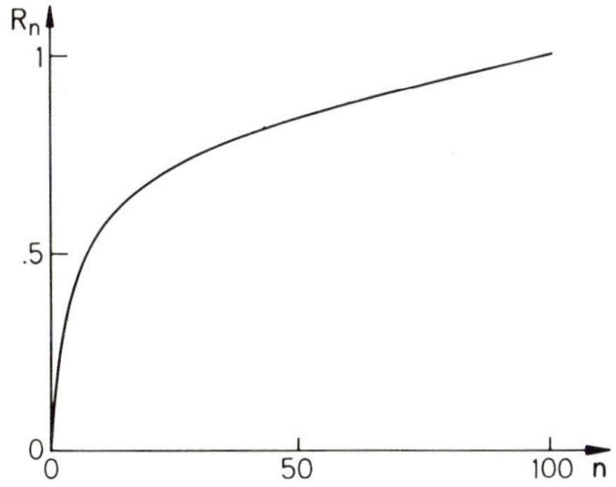

Fig. 6.9. Probability R_n of reaching the individual number K for a population of the present size (individual number) n for model in Eqs. (10) and (11) $R = 2r$, $\sigma^2 = 1.5r$ and $K = 100$

drawn from Figs. 6.4 and 6.5 is not valid for cases with strong environmental noise.

6.2.2.3 Dependence on Various Parameters

In the discussion of Fig. 6.2, we emphasized the importance of the strength σ of the environmental noise which is the second important quantity besides the carrying capacity K. A typical dependence of the mean life time T_K on σ is shown in Fig. 6.10. Varying $\sigma/\sqrt{r}$ from 0.1 to 1, the life time drops dramatically from $T = 10^9/r$ to $T = 10^2/r$. This is related to the fact that the mean life time

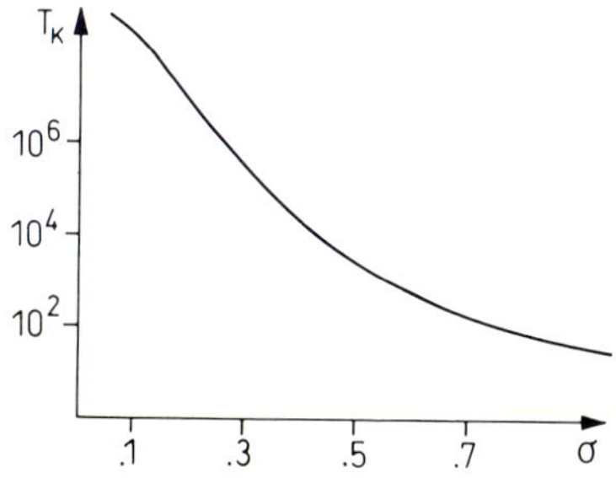

Fig. 6.10. Mean life time T_K (in units of 1/r) of a population versus the strength σ of the environmental noise (in units of $\sqrt{r}$) for the model in Eqs. (10) and (11) with $R = 1.1r$ and $K = 20$

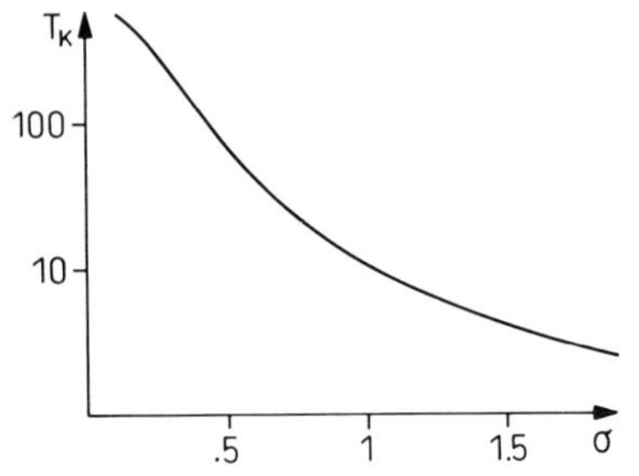

Fig. 6.11. Mean life time T_K (in units of 1/r) versus the strength σ of the environmental noise for the model in Eqs. (10) and (11) with R = 3r and K = 20

T_K increases strongly for increasing K for small σ [see Eq. (16)] and only moderately for large σ. Figure 6.11 shows that this dependence of the mean life time T_K on σ is not so dramatic if the demographic noise R is stronger. However, this influence of the strength of the demographic noise decreases for increasing values of the carrying capacity K. The explicit dependence of the mean life time T_K on the strength R of the demographic noise is shown in Fig. 6.12. This dependence is the weaker, the larger the carrying capacity K, and the larger the strength σ of the environmental noise are. Therefore, the sensitivity with respect to the one parameter depends strongly on the size of the other parameters. The dependence on K is the stronger, the smaller σ is. The dependence on one noise strength is the smaller, the larger the other noise strength is, and the larger K is. Especially the dependence on R becomes weak if K is large. Remember that we found from Fig. 6.7 that the mean life time can be used to obtain an estimate for MVP. A risk of extinction of 5% is obtained at a time of about 0.1 T_K.

So far, we have considered the logistic growth only. In this case, density dependence already starts at small individual numbers n and increases linearly. Another extreme case is the model [Eqs. (10a) and (11a)] where there is no density dependence at small individual numbers and where density regulation appears at n = K only. In Fig. 3, the corresponding dependence of the mean life time T_K on the carrying capacity K is shown. The life time is longer in this case. The difference to the life time of the logistic growth increases from a factor of 3 at small K to a factor of almost 100 at greater K (cf. Goodman 1987a). This difference is the smaller, the larger the strength σ of environmental noise is.

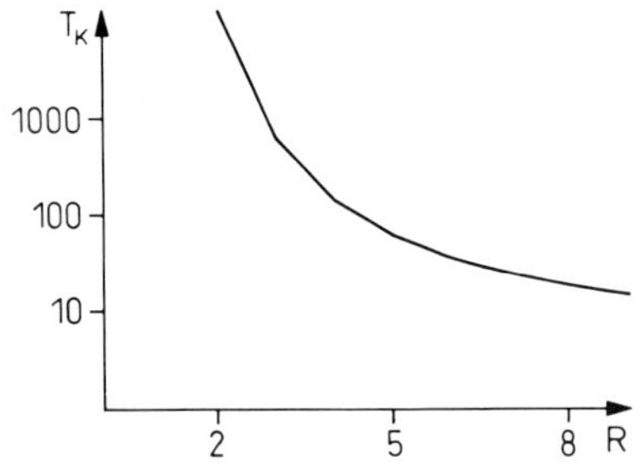

Fig. 6.12. Mean life time T_K (in units of 1/r) versus the strength R of the demographic noise for the model in Eqs. (10) and (11) with $\sigma^2 = 0.01r$ and K = 20

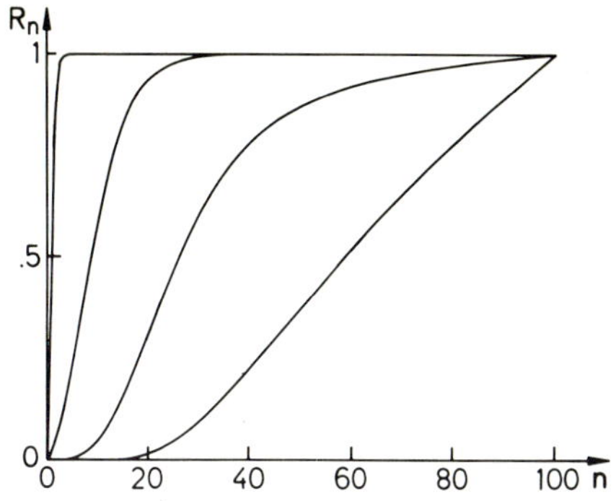

Fig. 6.13. Influence of the Allee effect on the probability R_n of reaching the individual number K for a population which contains n individuals at present. *From top down* r_0 (intrinsic growth rate for n = 0) is 1.5, −0.5, −1, −1.4 [model in Eqs. (10) and (11) with (17) for r, with c = 2, v = 20, R = 2, $\sigma^2 = 0.1$ and K = 100]

Another reason for a different density dependence at small individual number may be the Allee effect. It describes the decrease of the intrinsic growth rate r at a small individual number which may appear for various reasons (lack of cooperative support in predator defence or in breeding, reduced probability of finding mates etc.). As an example, we consider (Jacobs 1984)

$$r = r_0 + \frac{cn}{n + v}.$$

This means that the intrinsic growth rate r increases with n and levels off at higher values of n. The initial slope of increase is given by c/v and the saturation level is c. It is obvious that this decrease in r at small n must have a strong effect on the mean life time, as the range of small n is of special importance, as discussed above. The diminution of the life time by the Allee effect depends strongly on the parameter values and can be very strong.

Another important consequence of the Allee effect is the change of the probability R_n of reaching the individual number K for a population with n individuals at present. In Fig. 6.13 an example is shown for different values of r_0. The form of the curve is changed when decreasing r_0, i.e. the minimal growth rate for very small individual numbers. The critical value n_c, which is $n_c = 3$ without the Allee effect, increases considerably. It delimits the range of individual numbers where the population is relatively safe. A further decrease in r_0 results in a complete vanishing of this safer range. The population becomes much more vulnerable to disturbances which cause short-term decrease in individual number n down from K. For a colonizing population, the consequences are as expected. At very small n there is no chance R_n of successful colonization because of the negative intrinsic growth rate. Only if the number n of colonizators is above a certain number does the probability of success R_n increase from zero.

One may be interested in help for a population which is endangered by the risk of extinction. Let us consider the case in which there is always a chance that individuals immigrate. For high numbers of individuals, this small chance is negligible compared to the numerous birth processes; but at low individual numbers n with few birth processes, this additional increase in the individual

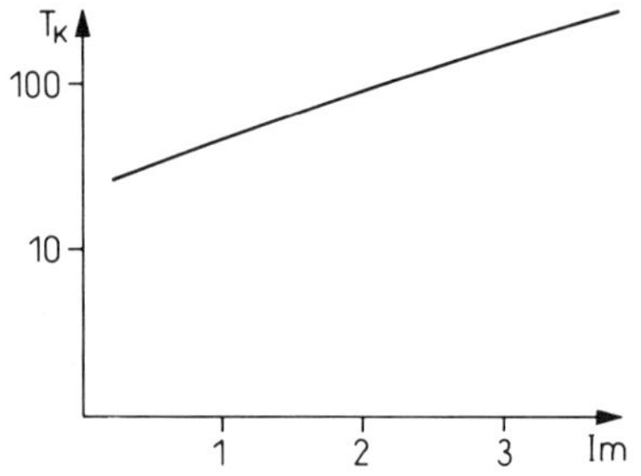

Fig. 6.14. Mean life time T_K (in units of 1/r) of an established population versus immigration rate *Im* (in units or r) for model in Eqs. (10) and (11) with $R = 1.1r$, $\sigma^2 = r$ and $K = 20$

number can be noticeable. As we know that the range of small numbers n plays an important role for extinction processes, we expect that a small immigration rate Im has an effect on the mean life time T_K. Immigration is an increase in the individual number, as the birth rate is. Therefore, immigration is described by adding a constant rate Im to λ_n

$$\lambda_n \rightarrow \lambda_n + \text{Im},$$

with Im being the probability of the arrival of an immigrant per time. In Fig. 6.14 the influence of immigration on the mean life time T_K is shown. A small immigration rate which equals the growth rate of only three individuals prolongs the life time by a factor of 10. There is another reason why a small immigration can be very helpful for the survival of a species; although a risk of extinction remains, there exists a chance of recolonization in the presence of immigration. Therefore, there may be a permanent change of existence and absence of a species in a particular habitat. However, the existence of the species in the long run is guaranteed by the immigration. This example shows that a longer existence of a species in a particular habitat does not mean that the habitat per se is sufficient for the survival of the species. If existing small immigration "streets" are cut, the species may die out in this habitat. On the other hand, the creation of small immigration by corridors or "stepping stones" can have great influence on the existence of the species.

Another possibility of supporting the survival of a population is the improvement of the habitat. This can be achieved by measures which diminish the mortality of adult fertile individuals. In Fig. 6.15, it is shown how a decrease in the death rate increases the mean life time T_K. Remember [see Eqs. (6) and (7)] that the individual death rate μ_1 is $(R - r)/2$, which is $\mu_1 = r$ for the present parameter values. Thus μ_1 is decreased by a factor of 3 in Fig. 6.15, which results in an increase of T_K by a factor 100. This means that reduction of mortality has a strong effect on the chance of survival of populations.

The improvement of a habitat may also increase the birth rate. Here we must remember that the birth rate λ_n provides the increase of the number n of adult, fertile individuals (see Appendix A). Therefore the juvenile mortality up to the fertile stage is included in λ_n. Consequently, an increase of this "birth rate" can be achieved also by a decrease in the juvenile mortality. In Fig. 6.16, the effect

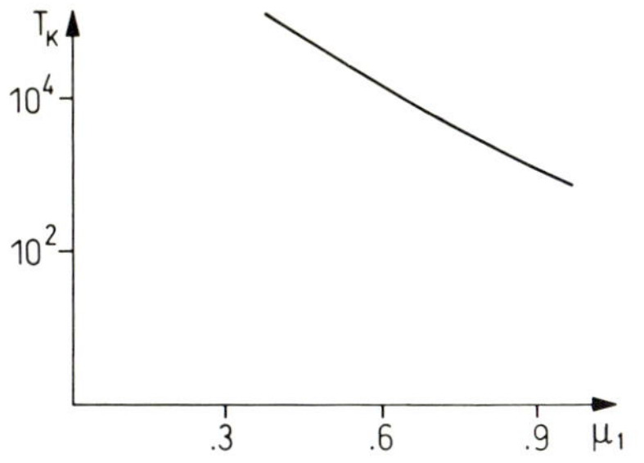

Fig. 6.15. Mean life time T_K (in units of 1/r) versus individual death rate μ_1 (in units of r) [see Eq. (5)] for model in Eqs. (10) and (11) with $\lambda_1 = 2r$, $\sigma^2 = 0.47r$ and $K = 100$

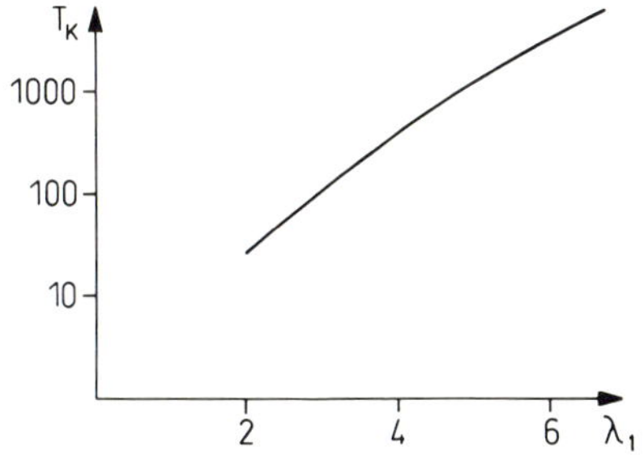

Fig. 6.16. Mean life time T_K (in units of 1/r) versus individual birth rate λ_1 [see Eq. (4)] (in units of r) for model in Eqs. (10) and (11) with $\mu_1 = r$, $\sigma^2 = 0.49r$ and $K = 20$

of an increase in the individual birth rate λ_1 on the mean life time is shown. Using Eqs. (6) and (7), the original individual birth rate λ_1 equals $(R + r)/2$, which is $\lambda_1 = 2r$ for the present parameter values. Therefore an increase in λ_1 by a factor of 3 prolongs the mean life time by a factor of 300. Generally, one can conclude that increase in birth rate or decrease in juvenile mortality is very effective in prolonging the life time of a population.

So far we have considered adult, fertile individuals with mean properties only. The explicit consideration of different sexes leads to models which are more difficult to solve. In this case, tedious numerical methods have to be used, which cannot be shown in this chapter. In the case of polygamy, we find almost no difference in the results compared to our model [Eq. (3)]. In the case of monogamy, we find values of the mean life time T_k which are shorter by a factor of about 10 compared to the values of model (3).

6.3 Population Dynamics with Non-Overlapping Generations

6.3.1 Time Discrete Modelling

Some species show a population dynamics with well separated generations. These may be annual plants or semelparous insects. Often their reproduction is synchronized with the annual cycle of the weather. For the description of this population dynamics in the course of years, one may choose the number

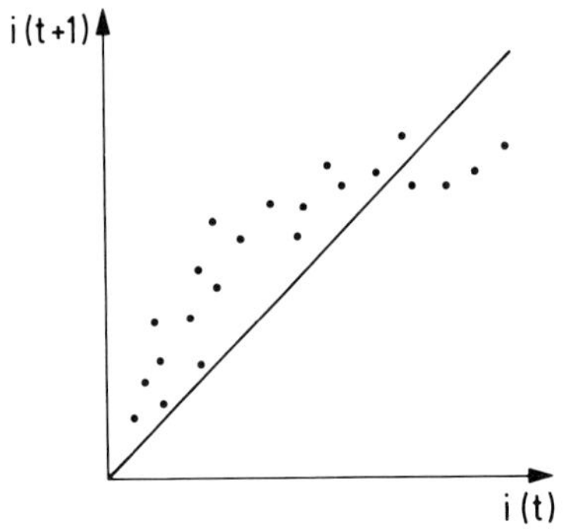

Fig. 6.17. General form of a population dynamics with non-overlapping generations. Individual number $i(t+1)$ in the following year $(t+1)$ versus number $i(t)$ in year t. At the *straight line* it is $i(t+1) = i(t)$

i(t) of fertile individuals in year t. The dynamics of the population is determined by the dependence of the individual number $i(t+1)$ in the next year on the number i(t) in this year. In Fig. 6.17, a typical form of the dependence of $i(t+1)$ versus i(t) is shown. Disregarding the spreading of the points, one may try to describe the general trend by fitting a curve on these points. As an example we take (Wissel 1989b)

$$i(t+1) = \lambda(i) = \frac{Ri}{1+ai}, \tag{19}$$

where $i = i(t)$. The meaning of R is the maximal reproduction rate (number of propagules in the next year), which is found for small individual numbers i where no density dependence appears. The parameter a describes the strength of density dependence and determines the equilibrium value K where population growth stops.

$$K = \frac{R-1}{a}. \tag{20}$$

The decisive point for the further discussion is the fact that the points in Fig. 6.17 do not follow a deterministic relation like Eq. (19), but show a considerable spreading. This means that Eq. (19) describes only the mean behaviour, and deviations from this curve appear by chance. The only possibility of dealing with this random process is the description by probabilities. Therefore we introduce the probability A_{ki} that the individual number is k in the next year if it is i in this year. This transition matrix A_{ki} determines the whole stochastic population dynamics. Because of the random nature of the population growth, we can describe the individual number in the course of time by probabilities only. Let $P_i(t)$ be the probability that the individual number is i in the year t. Then

$$P_k(t+1) = \sum_i A_{ki} P_i(t) \tag{21}$$

is the equation which describes the stochastic population dynamics from year to year. All possible information is contained in $P_i(t)$. Equation (21) is called a Markov chain.

If there are enough data points in Fig. 6.17 from field investigations, one may determine the transition matrix A_{ki} by fitting the parameters of a corresponding formula to the data. In this way one obtains a phenomenological description without any reference to the underlying biological processes. In the theoretical literature (see Gabriel and Bürger 1992), a Poisson distribution has been taken for A_{ki}

$$A_{ki} = \frac{\lambda(i)^k}{k!} e^{-\lambda(i)}, \tag{22}$$

with the mean value $\lambda(i)$ given by Eq. (19). This distribution has the advantage that it needs no further parameter. Another choice for A_{ki} with the same mean value $\lambda(i)$ is a normal distribution

$$A_{ki} \sim \exp\left[-\frac{1}{2}\left(\frac{1 - k/\lambda(i)}{\beta}\right)^2\right]. \tag{23}$$

The poisson distribution [Eq. (22)] and the normal distribution [Eq. (23)] differ by the standard deviation Sd(k).

$$\begin{aligned} \mathrm{Sd}(k) &= \sqrt{\lambda(i)} \quad \text{Poisson} \\ \mathrm{Sd}(k) &= \beta\lambda(i) \quad \text{normal.} \end{aligned} \tag{24}$$

Which form is the better description can be decided by the fit to the data points in Fig. 6.17 only.

Another possibility is the modelling of the biological processes which determine the transition probability A_{ki}. Depending on the specific ecological situation, there may be many different possible models. Here we represent the modelling of some examples which exemplify some general trends.

Let us consider the modification of Eqs. (19) and (22), which results if sexual reproduction is taken into account. As only females can give birth, we have to write instead of Eq. (19) for the mean number of propagules (Gabriel and Bürger 1992)

$$\lambda(i) = \frac{2Rj}{1 + ai}, \tag{25}$$

where j is the number of females. Equation (25) gives the same mean growth rate as in Eq. (19) because the mean total number (females and males), i equals 2j if the mean sex ratio is 1:1. The denominator remains unchanged as males and females contribute to the density regulation by intraspecific competition. Equation (25) is valid only as long as there is at least one male in the population ($j < i$). This would describe polygamous mating. For monogamous mating we have to substitute the number of pairs for j.

If sex is determined by chance, the probability $W_i(j)$ of finding j females in between i individuals is gives by a binomial distribution (see Appendix E). There is no need to write an equation different from Eq. (21) (Gabriel and Bürger 1992), which contains the number of females explicitly. Appendix E shows that one can reduce the description of the population dynamics to Eq. (21) also in this case. Instead of Eq. (22), another transition matrix A_{ki} is found, which contains the information of sexual reproduction. The only point of importance for the population dynamics is the fact that this matrix A_{ki} shows a much larger standard deviation for the transition probabilities. This means that the spreading of the data points in Fig. 6.17 would be increased by sexual reproduction.

Next let us consider a more explicit model with different stages of development of individuals. As an example, we take an insect population. Let us suppose that each fertilized female produces 2R eggs on average, so that the total number of eggs is on average

$$E = 2Rj. \tag{26}$$

The females may be exposed to changing weather conditions and, consequently, the number of eggs may vary. As all individuals are influenced by good or bad weather, the variation of the number of eggs per female is independent of how many individuals (eggs) are present. This means that we model environmental noise in the same way as in Eq. (2). Assuming a normal distribution for the action of weather, we write for the probability of obtaining l eggs in the population.

$$W_E(l) \sim \exp\left[-\frac{1}{2}\left(\frac{1-l/E}{s}\right)^2\right]. \tag{27}$$

Here s gives the relative standard deviation (standard deviation divided by the mean E).

Then let us assume that in the larval stage intraspecific competition appears which gives rise to density-dependent regulation. That means some of the larvae die by competition. Let $u(l)$ be the probability that a single larva survives and develops to an imago in the next year if one starts from l eggs. We take

$$u(l) = \frac{f}{(1 + al)^{\gamma}}, \tag{28}$$

with f being the density-independent survivership (at small l) for one larva and a being the strength of density-dependent regulation. Here we have introduced an additional parameter γ which describes contest competition (Wissel 1989b) for $\gamma = 1$ and scramble competition for large γ. With this survival probability of a single larva, the survival of all larva is described by a binomial distribution (see Appendix E).

Finally, an Allee effect may appear, i.e. a reduced reproduction at small individual numbers. We consider the situation that the chance of finding a mate is reduced at small individual numbers. Let w_0 be the probability that one particular male finds one particular female. The probability that one particular female will be found by any one of $(i-j)$ males (polygamy) and consequently will be fertilized is

$$W(i,j) = 1 - (1 - w_0)^{i-j}. \tag{29}$$

Using this for the fertilization probability of a single female, the total number of fertilized females is described by a binomial distribution (see Appendix E). Thus, instead of the number j of females we have to use this number of fertilized females in Eq. (25).

It should be emphasized that all these details of the model are finally combined in the transition matrix A_{ki} (see Appendix E). Only this matrix determines the population dynamics. We have given only one possible example for modelling the population dynamics, more detailed in order to demonstrate that such modelling is possible in principle. This model will be used to exemplify some general trends. The other possibility is a phenomenological model [see Eq. (22) or Eq. (23)] without any modelling of biological details.

6.3.2 Results

The quantities of interest which can be calculated are exactly the same as in Section 6.2.2. How they can be calculated in this case is shown in Appendix E. As a standard set of parameter values we take (f = 1) (no death of larvae without density dependence), $\gamma = 1$ (contest competition), s = 0 (no environmental noise) and W = 1 (no Allee effect) for our specific model, Eqs. (26) to (28), as long as nothing else is said.

In Fig. 6.18, the probability R_n of reaching the established state at the carrying capacity for a population of n individuals is shown. In Fig. 6.19, the mean life time T_n of a population which contains n individuals at present is shown versus n. The consequence of these typical curves is the same as in Section 6.2.2. There is a critical value n_c for the individual number, which delimits a range of reduced risk of extinction. An approximate formula for R_n valid for great R and K can be given by

$$R_n \approx 1 - (\tfrac{1}{2})^{n-1}. \qquad (30)$$

The interpretation is obvious. The second term of Eq. (30) is the probability that an individual finds other individuals (n − 1) of the same sex only. Thus

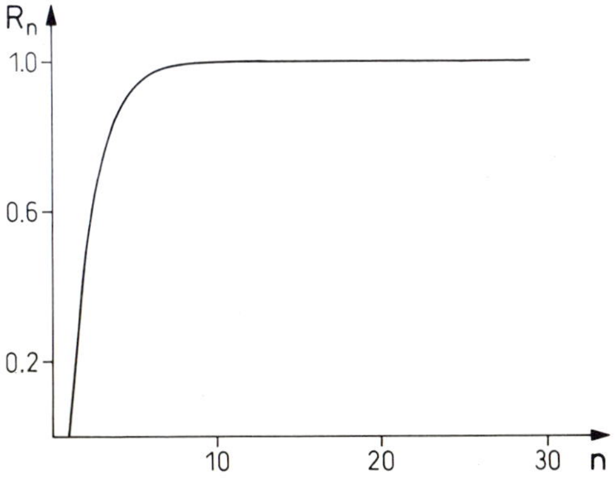

Fig. 6.18. Probability R_n that the stage of an established population (carrying capacity K) is reached if the present individual number is n. [Model in Eqs. (26) to (28), carrying capacity K = 30, maximal reproduction rate R = 10, polygamy]

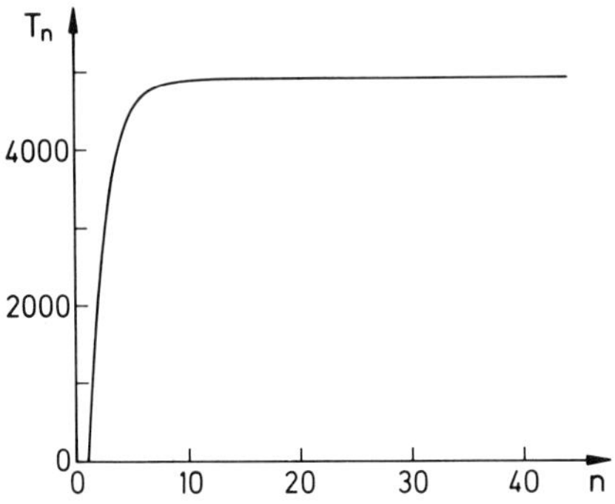

Fig. 6.19. Mean life time T_n (in years) of a population which contains n individuals at present. [Model in Eqs. (26) to (28), K = 20, R = 3, polygamy]

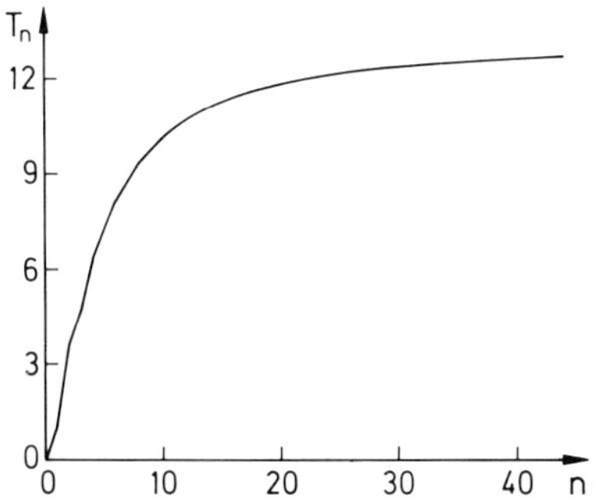

Fig. 6.20. Mean life time T_n (in years) of a population which contains n individuals at present [Model in Eq. (22), K = 15, R = 2, monogamy]

formula (30) is the probability that at least one pair is present in between n individuals. This is obviously sufficient to guarantee the establishment of a population at the carrying capacity if the reproduction rate R is sufficiently large. Therefore, the stochasticity which results from the presence of two sexes is most important (here no environmental noise). If n_c is defined by $R_{n_c} \geqslant 0.95$, we find from Eq. (30) that $n_c = 6$. The same results are obtained for the phenomenological model (22) if the carrying capacity K and the maximal reproduction R are sufficiently high. For smaller values of R and K, the value of n_c is somewhat enlarged.

However, at very low values of R and K, no saturation behaviour like that in Figs. 6.18 and 6.19 is found, an effect which is more pronounced in the monogamous model. Figure 6.20 for the mean life time and Fig. 6.21 for the probability of establishment exemplify this behaviour.

In interpreting Fig. 6.19, we cannot argue as in Section 6.2.2, because here the mean life time T_n is not additive. Therefore in Table 6.2, the mean life times T_K and T_5 and the mean time T_{5K} to reach the individual number n = 5 from the carrying capacity K are shown. As in Section 6.2, a population needs less time to come down to small individual numbers compared to the time the population needs to become extinct from these small numbers. Therefore the details of the model for high individual numbers are not important. This has

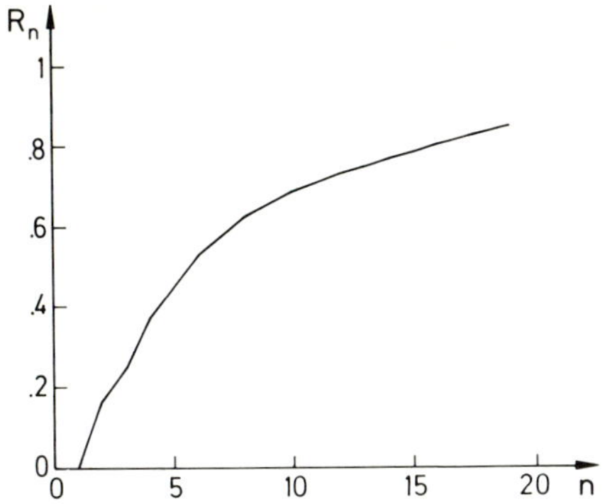

Fig. 6.21. Probability R_n of reaching the established state at K for a population with n individuals at present. [Model in Eq. (22), K = 20, R = 2, monogamy]

Table 6.2. Mean life time of a population with n = K and n = 5 individuals and mean time T_{5K} a population needs for lowering the individual number from K down to n = 5 [model in Eqs. (26) to (28), maximal reproduction rate R = 3 and K = 20]

	T_K	T_5	T_{5K}
Polygamy	4900	4490	1410
Monogamy	1040	913	278

Table 6.3. Dependence of the mean life time T_K on carrying capacity K and maximal reproduction rate R for model in Eq. (22) with monogamy

R/K	15	20	25	40
3	17.1	38.9	98	2730
10	36.8	140	598	6.1×10^4
1000	49.3	216	1030	1.43×10^5

been checked by modifying the model at high numbers. On the other hand, the details at smaller numbers are important. There the standard deviation of the transition matrix A_{ki} is decisive.

The dependence of the mean life time T_K on the carrying capacity K and the maximal reproduction rate R is shown in Tables 6.3 and 6.4 for the model in Eqs. (22) and (26)–(28), respectively. The increase in T_K with growing K is exponential. This is the same behaviour which is found in the model in Eqs. (10) and (11) for overlapping generations without environmental noise, i.e. $\sigma = 0$. Here, also no environmental noise is taken into account so far. The use of the Poisson distribution in model (22) means that the distribution results from many independent events. This causes the standard deviation to increase only with the square root of the mean. That this corresponds to what we called demographic

Table 6.4. Dependence of the mean life time T_K on carry capacity K and maximal reproduction rate R for model in Eqs. (26) to (28) with polygamy

R/K	15	20	25	40
3	484	4900	5.27×10^4	7.60×10^7
10	914	11 800	1.54×10^5	3.27×10^8
1000	905	11 000	1.34×10^5	2.43×10^8

noise in Section 6.2 becomes more obvious for the model in Eqs. (26) to (28). There, the individual variability means that only a certain probability u(l) for the survival of a single individual can be given. For the total number of surviving individuals, it is assumed that the fates of different individuals are independent, which results in a binomial distribution. This distribution also shows an increase in the standard deviation proportional to the square root of the mean. This moderate increase is responsible for the exponential growth of the mean life time T_K with increasing carrying capacity K. The inclusion of different sexes in the model (an individual variability) leads also to a binomial distribution for the number of females and consequently contributes to the demographic noise. In the model in Eqs. (26) to (28) we use binomial distributions several times, which all add to demographic noise. Environmental noise is discussed later.

In Section 6.2 we saw that the probability $P_0(t)$ that the population is extinct at time t can be used to determine minimum viable populations (MVP). A typical result is shown in Fig. 6.22. For large times, extinction is almost sure [$P_0(t) = 1$]. An extinction probability of $P_0(t) = 5\%$ is typically reached at a time which is smaller than the mean life time T_K by a factor of about 20. With this in mind, values of T_K can be used for assessing MVPs.

After a short initial phase $P_0(t)$ takes the form

$$P_0(t) = 1 - c\,e^{-\omega t}. \tag{31}$$

In Appendix E it is shown that ω equals $1/T_K$. Therefore plotting $-\ln[1 - P_0(t)]$

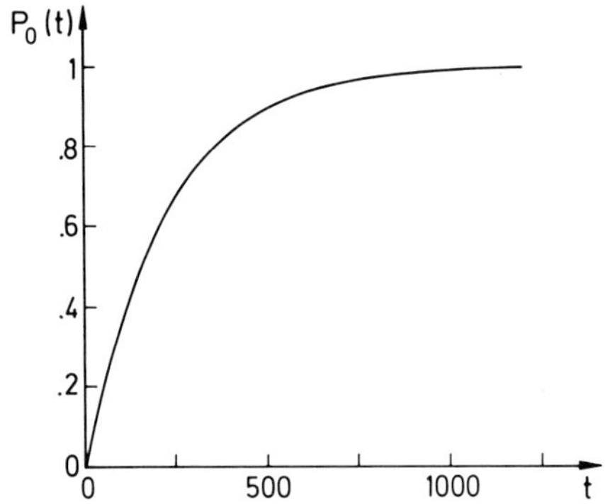

Fig. 6.22. Probability $P_0(t)$ that the population is extinct at time t for model in Eq. (22) with K = 20, R = 3 and polygamy. The mean life time is $T_K = 224$ years

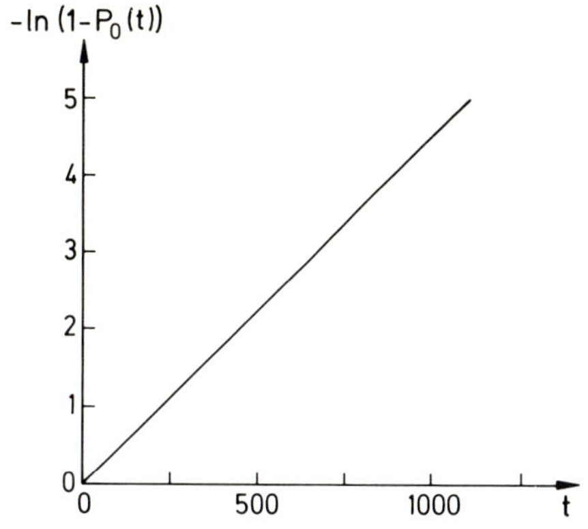

Fig. 6.23. Probability $P_0(t)$ as in Fig. 6.22, here using the scale $-\ln[1 - P_0(t)]$

Table 6.5. Influence of different model modifications on the mean life time T_K (in units of the mean life time of model in Eq. (22) with monogamy; $K = 20$ and $R = 10$

Model type		T_K
Model (22)	Monogamous	1
(Poisson)	Polygamous	3.6
	Asexual	20 500
Model (26)–(28)	Monogamous	6.7
	Polygamous	8.5
	Asexual	9.2×10^6

versus t, as shown in Fig. 6.23, one obtains a straight line with the slope ω. This is another way [cf. Eq. (E9) in Appendix E] of determining the mean life time $T_K = 1/\omega$.

In order to assess the influence of the different model modifications, an example is given in Table 6.5. We see that monogamy gives the shortest mean life times. Polygamy increases T_K, but not more than by one order of magnitude. Asexual reproduction increases T_K by several orders of magnitude.

So far we used $\gamma = 1$ in Eq. (28), which means contest competition (Wissel 1989b). Now let us consider scramble competition by choosing $\gamma = 10$. The corresponding mean number of propagules versus the number of individual is shown in Fig. 6.24. Scramble competition causes higher individual numbers to give fewer propagules in the next year. In Table 6.6, corresponding mean life times are shown, which demonstrate that scramble competition leads to short life times if reproduction rate R is large.

Instead of the phenomenological model in Eq. (22) with a Poisson distribution, one may consider model (23) with a normal distribution. In Fig. 6.25, the relative standard deviation of both distributions is compared. The Poisson distribution shows a larger relative standard deviation at small individual numbers. As the range of small individual numbers determines the mean life time to a large

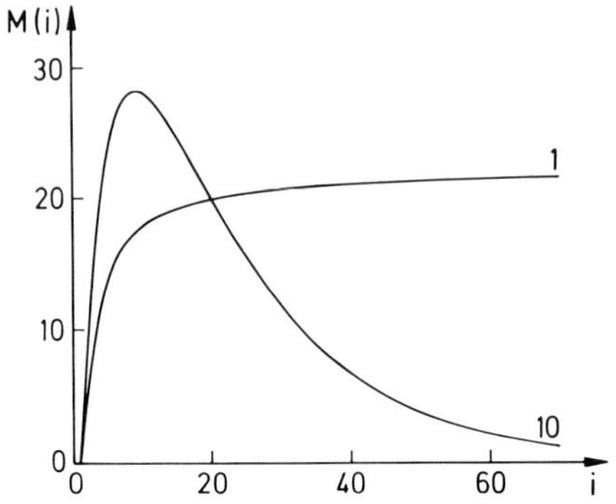

Fig. 6.24. Mean number $M(i)$ of propagules versus number i of individuals for ($\gamma = 1$) contest and ($\gamma = 10$) scramble competition for model in Eqs. (26) and (28) with K = 20 and R = 10

Table 6.6. Mean life times T_K for scramble ($\gamma = 10$) and contest ($\gamma = 1$) competition for K = 20

Polygamous model (26, (28)			Monogamous model (20)	
R	contest ($\gamma = 1$)	scramble ($\gamma = 10$)	contest ($\gamma = 1$)	scramble ($\gamma = 10$)
3	4900	9860	38.9	75.9
10	11 800	146	990	96

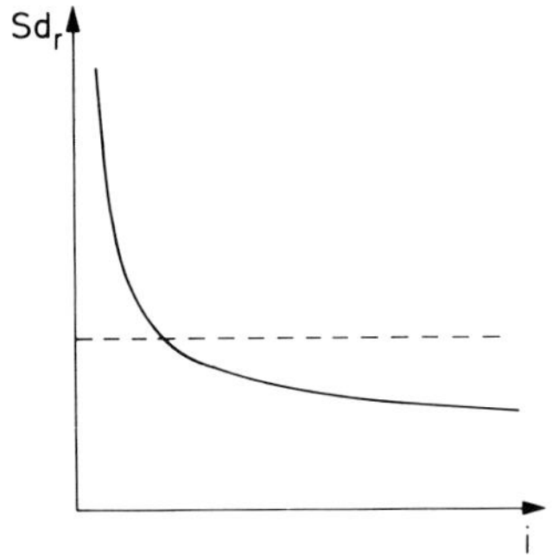

Fig. 6.25. Relative standard deviation (*Sdr*: standard deviation divided by the mean) of the Poisson [Eq. (22)] distribution (*full line*) and of the normal [Eq. (23)] distribution (*dashed line*) versus individual number i

extent, we expect the model in Eq. (22) with the Poisson distribution to show shorter life times than the model in Eq. (23) with the normal distribution. This indeed is found.

Next, we consider the influence of the Allee effect as described in Eq. (29). As long as the probability w_0 that one particular female is found by one particular male is near $w_0 = 1$, little effect on the mean life time T_K is seen. The Allee effect shows a stronger influence on T_K only for smaller values of w_0. An example is shown in Table 6.7. It is seen that the mean life times T_K can be considerably shortened by the Allee effect, and that the life time T_5 when five individuals are reached is drastically reduced compared to the time to reach five individuals from the capacity K.

Table 6.7. Mean life time T_K and T_5 and mean time for reaching $n = 5$ from $n = K$ for Allee effect for model in Eqs. (22) and (29) with $K = 20$, $R = 3$ and polygamy. Parameter w_0 is the probability that one particular female is found by one particular male

w_0	T_K	T_5	T_{5K}
0.1	7.0	1.8	6.0
0.2	57.7	15.8	48.1
0.3	530	296	326
0.4	1790	1330	816
1	4900	4480	1410

Table 6.8. Mean life time dependent on the density-independent survival rate f and the reproduction rate R for model in Eq. (20) with $K = 20$ and polygamy

R	f	T
3	1	4900
30	0.1	1205
10	1	11 800
100	0.1	8450

So far we considered $f = 1$ in Fig. (28), which means that there is no mortality without density dependence of the larvae. Let us consider the case in which f, the density-independent survival rate, is reduced by a factor of 10, but the mean reproduction rate R is increased by the same factor. This makes no differences as long as mean values are considered. However, by this change, the standard deviation of the total numbers of survivors is increased, which results in a decrease in the mean life time T_K. In Table 6.8, it is shown that this effect is stronger at small values of the reproduction rate R.

Finally, we have to consider the effect of environmental noise. It is described in Eq. (27) and influences the number of eggs produced by one female. Assuming that females lay eggs independently from each other would result in a binomial distribution and would consequently contribute to demographic stochasticity; but if we consider changing weather as an example for environmental noise, it is obvious that all females suffer from bad weather conditions and may reduce the number of their eggs simultaneously. In contrast to demographic noise, environmental noise acts on the individuals in a correlated manner. The relative standard deviation s of the distribution is a measure for the strength of environmental noise. In Table 6.9, it is shown that environmental noise can reduce the mean life time T_K by several orders of magnitude. A relative standard deviation of $s = 0.3$ already diminished T_K considerably. This is in complete agreement with the results of Section 6.2 for overlapping generations. In

Table 6.9. Mean life time T_K for different strength for environmental noise s for model in Eq. (22) with K = 20, R = 3 and monogamy

S	0	0.2	0.3	0.3	1
T_K	1040	547	190	36.4	17.5

Table 6.10. Dependence of the mean life time T_K on reproduction rate R and carrying capacity K for model in Eq. (22) with polygamy and strength of environmental noise s = 0.5

R/K	15	20	25	30
3	39.6	64.7	90.3	116
10	153.9	300	431	553

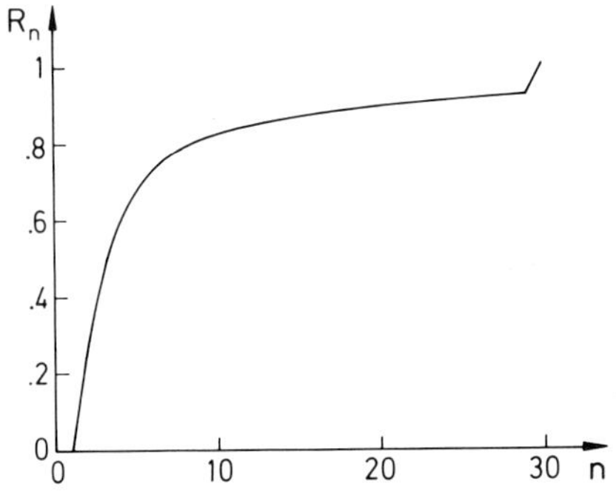

Fig. 6.26. Probability R_n of reaching the established state at K for a population with n individuals at present for model in Eq. (22) with K = 30, R = 3, polygamy and strength of environmental noise S = 1

Table 6.10, the dependence of the mean life time T_K on the reproduction rate R and the carrying capacity K is shown. Now the increase in T_K with growing K is much smaller than in the case without environmental noise (see Tables 6.3 and 6.4), which agrees with the findings of Section 6.2 for overlapping generations. Finally, stronger environmental noise changes the form of the probability R_n that a population with n individuals reaches the established state at the carrying capacity K. An example is given in Fig. 6.26 which shows that there is no saturation behaviour with a critical individual number n_c as in Fig. 6.18. Therefore, there is no range of individual numbers with reduced risk of extinction if a stronger environmental noise is present. This agrees completely with the results for overlapping generations in Section 6.2.

6.4 Conclusion

For the survival or extinction of small populations, only probabilistic statements can be made. Theoretical ecology (mathematical models) is able to reveal some general trends which are important for the assessment of the risk of extinction. When modelling small populations, one should distinguish between species with overlapping generations and species with separated generations. For the first case, one has to model the birth and death rates and in the second case, a transition probability that a certain number of individuals gives rise to a certain number of propagules has to be given. For both cases, new calculation techniques are given in the Appendices. In both cases, density dependence has to be modelled. Stochasticity which arises from individual variability is called demographic noise. It is of special importance at small individual numbers, whereas its influence decreases at high numbers. Environmental noise acts on all individuals simultaneously, whereas demographic noise acts on individuals independently. Environmental noise is more important at high individual numbers. It causes occasionally small individual numbers to be reached. There, demographic noise becomes important and gives rise to a finite chance of extinction. Demographic noise is determined by the basic demographic parameters like birth rates and mortalities. The strength of environmental noise is more difficult to assess. Under environmental noise, all temporal varying influences which cannot explicitly be taken into account are subsumed. For assessing the strength of environmental noise, the relative standard deviation for the change of the individual number during a certain time interval has to be known.

We have found closed formulae for the mean life time T_K of a population with n individuals at present and the probability R_n that this population reaches the established state at the carrying capacity. If environmental noise is not too strong, a saturation behaviour is found for T_n and R_n. This means that there is a critical number n_c which delimits a range of individual numbers which is relatively safe from extinction. Populations with individual numbers above n_c return almost surely to the carrying capacity. Below n_c there is an enlarged risk that the population becomes extinct, and the mean life time of this population is reduced. Therefore n_c is the minimal individual number to which a population can be reduced by some disturbance without an enhanced risk of extinction. The range below n_c gives the critical stage of higher risk of extinction which a colonizing population has to pass. Therefore there are two essential individual numbers of a population, the carrying capacity K (deterministic equilibrium), which is determined by the amount of limiting resources of a habitat, e.g. size of a habitat, and the critical number n_c, which delimits the safer range and is determined by demographic parameters. It has to be emphasized, however, that for strong environmental noise, a critical number n_c does not exist. There is no safer range in this case.

Often the influence of the size of a habitat on the chance of survival is emphasized. For smaller environmental noise there is indeed a strong dependence of the life time on the carrying capacity. In this case, the enlargement of a habitat is a good strategy to support the chance of survival of an endangered population; but the stronger the environmental noise, the weaker the dependence of the life time on the capacity K. Therefore in the case of strong environmental noise, the increase in the carrying capacity does not produce the wanted effect. In this case, other strategies which are mentioned below are more effective.

Mean life times can be used if one wishes to determine minimum viable population (MVP). It is defined as the smallest population with a risk of extinction which does not exceed a given limit during a given period of time. A risk of extinction always remains. The decisive question which cannot be answered by science is which risk of extinction for which period of time one is willing to tolerate. This is a question of politics. One can deal with MVP in two different ways. Given a habitat which provides a certain carrying capacity, one may ask for the time T′ when the probability of extinction reaches the 5% limit. Our models show that this time T′ is about one order of magnitude shorter than the mean life time T_K of an established population. Therefore, the calculation of T_K can be used to determine this limiting time T′. On the other hand, one may prescribe a 5% limit of the risk for a given period of time (e.g. 100 years) and may ask how large the carrying capacity must be in order to prevent a risk higher than 5%. This can be assessed by the inverse relation between mean life time T_K and carrying capacity using the relation between T and T′.

A general trend of all models is that the time a population needs to come down to smaller individual numbers is short compared to the time of extinction starting from these small individual numbers. This difference is not so pronounced for strong environmental noise. This means that fluctuations around the carrying capacity are fast. The final decay of a population from small numbers needs a longer time on average because the population may recover. Therefore, details of population dynamics at smaller individual numbers are relatively more important. For instance, an Allee effect which is effective at small numbers can shorten the life time considerably.

The consideration of different sexes is especially important at small individual numbers. There, the chance that all individuals are of the same sex is rather large. This fact is especially important for species with non-overlapping generations. An individual has only one chance to find a mate, or it will not reproduce at all. For overlapping generations, an individual may find a mate later.

The fact that population dynamics at small individual numbers is decisive for the risk of extinction can be used to support an endangered population. A small immigration rate which is negligible at high individual numbers has a strong effect at small numbers. This is of special importance for strong environmental noise. There, an increase of the carrying capacity is of little use. Immigration is an alternative strategy to increase the chance of survival. For this case of strong environmental noise the creation of "immigration streets" or "stepping stones" can be very helpful. Our models show that another strategy is the improvement of the habitat, by which the influence of environmental noise is reduced, adult or juvenile mortality is decreased or birth rates are increased.

Finally, we have to warn that the precise numbers of the results cannot be taken too seriously. Our models can give only a very rough description. There are so many details which can be important, especially at small individual number. No model can take them all into account because all models are based on assumptions and idealizations; they can never provide exact numbers. Models cannot give recipes for nature conservation, but can reveal general trends, as shown above, which may be useful for making better decisions in conservational management.

Acknowledgements. A part (Sect. 6.3) of this research was supported by the Deutsche Forschungsgemeinschaft.

Appendix A

Given the data of age-specific survival rate l_x, i.e. the probability of reaching the age class x and the age-specific fecundity β_x of a species, the birth rate λ_1 and death rate μ_1 of an individual can be calculated in the following way:

As n describes the number of adult fertile individuals, one has to determine the age a, where an individual enters the fertile stage. The mean expected life span of an individual of age a is given by (Wissel 1989b)

$$e_a = \sum_{x \geqslant \alpha + 1} l_x/l_a. \tag{A1}$$

This value, which results from empirical life table data, has to be taken for the mean life time of model in Eq. (5) which is $1/\mu_1$.

The mean number of propagules (individuals older than age a) of an individual of the age a is (Wissel 1989b)

$$\upsilon_a = l_a \sum_{x \geqslant a} \beta_x l_x/l_a. \tag{A2}$$

This value obtained from empirical data has to be used in the model in Eqs. (4) and (5). There, the mean number of births by an adult individual is

$$\upsilon_a = \int_0^\infty e^{-\mu_1 t}\lambda_1 \, dt = \lambda_1/\mu_1. \tag{A3}$$

This has to be used to determine λ_1 with the help of Eq. (A2) if μ_1 is known.

Appendix B

Using the arguments of Goodman (1987a, b), one can consider the birth and death process of the model in Eq. (3) for a short time interval Δt as a combination of two Poisson processes. The first is determined by the birth rate λ_n. The mean value of the increase Δn of the individual number is

$$m_\lambda(\Delta n) = \lambda_n \Delta t. \tag{B1}$$

Its variance is

$$\upsilon_\lambda(\Delta n) = \lambda_n \Delta t. \tag{B2}$$

For the second process determined by the death rate μ_n, we obtain

$$m_\mu(\Delta n) = -\mu_n \Delta t \tag{B3}$$

$$\upsilon_\mu(\Delta n) = \mu_n \Delta t. \tag{B4}$$

Taking the sum of both processes, which are independent of each other (approximately for small Δt), we obtain for the mean increase Δn of the individual number:

$$m(\Delta n) = (\lambda_n - \mu_n)\Delta t, \tag{B5}$$

and for its variance:

$$\upsilon(\Delta n) = (\lambda_n + \mu_n)\Delta t,$$

which gives the standard deviation

$$\mathrm{Sd}(\Delta n) = \sqrt{\lambda_n + \mu_n}\sqrt{\Delta t}. \tag{B6}$$

For

$$\lambda_n = \tfrac{1}{2}[rn + \sigma^2 n^2]$$

$$\mu_n = \tfrac{1}{2}[-rn + \sigma^2 n^2],$$

we obtain

$$\frac{\mathrm{Sd}(\Delta n)}{m(\Delta n)} = \frac{\sigma n\sqrt{\Delta t}}{rn\Delta t}. \tag{B7}$$

Appendix C

The mean life time T_n of a population is given by the mean first passage time M_{on} (Goel and Richter-Dyn 1974) for the transition from n to 0.

$$M_{on} = \sum_{i=1}^{n} \sum_{S=i}^{\infty} (\mu_S \pi_{iS-1})^{-1} \tag{C1}$$

with

$$\pi_{ij} = \prod_{r=i}^{j} \mu_r/\lambda_r. \tag{C2}$$

The probability $R_{n,m}(0)$ of reaching the value n before the value 0 is reached, starting at the value m is given (Goel and Richter-Dyn (1974) by

$$R_{n,m}(0) = \sum_{i=0}^{m-1} \pi_{1,i} \Big/ \sum_{i=0}^{n-1} \pi_{1,i}. \tag{C3}$$

Especially we consider $R_n = R_{K,n}(0)$ and $W_n = R_{n,1}(0)$.

The *asymptotic dependence* of T_K on K can be calculated in the following way: instead of Eq. (C1) we can write:

$$M_{0n} = \sum_{i=1}^{n} \sum_{S=i}^{K} \frac{1}{\mu_S} \prod_{l=i}^{S-1} \frac{\lambda_l}{\mu_l} + \left(\sum_{i=1}^{n} \prod_{l=i}^{K} \frac{\lambda_l}{\mu_l} \right) \sum_{S=K+1}^{\infty} \frac{1}{\mu_S} \prod_{l=K+1}^{S-1} \frac{\lambda_l}{\mu_l}. \tag{C4}$$

Estimate of product:

$$\prod_{l=i}^{S-1} \frac{\lambda_l}{\mu_l} = \exp\left(\sum_{l=i}^{S-1} \ln \frac{\lambda_l}{\mu_l} \right). \tag{C5}$$

Since $\ln \lambda_l/\mu_l$ is a monotonously decreasing function of l for the model in Eqs. (10), (11) and Eqs. (10a), (11a) we obtain

$$0 \leqslant \sum_{l=i}^{S-1} \ln \frac{\lambda_l}{\mu_l} - \int_i^S \ln \frac{\lambda_l}{\mu_l} dl \leqslant \ln \frac{\lambda_i}{\mu_i} - \ln \frac{\lambda_K}{\mu_K} \leqslant \ln \frac{\lambda_1}{\mu_1}.$$

Substituting the sum in Eq. (C5) by an integral, the error is bounded by a constant. The integral gives for model in Eqs. (10a) and (10b):

$$\int \ln \frac{\lambda_n}{\mu_n} dn = \frac{2r}{\sigma^2} \ln n + n \ln \left(1 + \frac{2r}{\sigma^2 n + R - r} \right)$$
$$+ \frac{R+r}{\sigma^2} \ln \left(1 + \frac{R+r}{\sigma^2 n} \right) - \frac{R-r}{\sigma^2} \ln \left(1 + \frac{R-r}{\sigma^2 n} \right). \tag{C6}$$

The last three terms on the right side are bounded by a constant. Therefore we obtain:

$$\prod_{l=i}^{S-1} \frac{\lambda_l}{\mu_l} = \left(\frac{S}{i} \right)^{2r/\sigma^2} \cdot c' \tag{C7}$$

with a constant c′ independent of S and i. For the model in Eqs. (10), (11) the integral gives

$$\int \ln \frac{\lambda_n}{\mu_n} dn = 2r \frac{\sigma^2 + \frac{R}{K}}{\sigma^4 - \frac{r^2}{K^2}} \ln n + n \ln \frac{\lambda_n}{\mu_n} + \frac{R+r}{\sigma^2 - \frac{r}{K}} \ln \left(1 + \frac{R+r}{\sigma^2 - \frac{r}{K}} \frac{1}{n} \right)$$
$$- \frac{R-r}{\sigma^2 + \frac{r}{K}} \ln \left(1 + \frac{R-r}{\sigma^2 + \frac{r}{K}} \frac{1}{n} \right).$$

The last three terms of the right side are bounded by a constant. The first tends to $2r/\sigma^2 \ln n$ for $K \to \infty$. Therefore we obtain the same result as Eq. (C7). From this we find:

$$\frac{1}{\mu_S} \prod_{l=i}^{S-1} \frac{\lambda_l}{\mu_l} \to S^{\alpha-2} i^{-\alpha} c \tag{C8}$$

with

$$\alpha = 2r/\sigma^2. \tag{C9}$$

For an estimate of the first sum in Eq. (C4), we consider

$$\sum_{S=i}^{K} S^{\alpha-2} \geqslant \int_i^K S^{\alpha-2}\, dS = \frac{K^{\alpha-1} - i^{\alpha-1}}{\alpha - 1}, \tag{C10}$$

where the monotony of the function $S^{\alpha-2}$ is used. On the other hand, it is

$$\sum_{S=i}^{K} S^{\alpha-2} \leqslant i^{\alpha-2} + \int_i^{K+1} S^{\alpha-2}\, dS = i^{\alpha-2} + \frac{(K+1)^{\alpha-1} - i^{\alpha-1}}{\alpha - 1}. \tag{C11}$$

Therefore we obtain for $\alpha > 1$:

$$\sum_{i=1}^{n} \sum_{S=i}^{K} S^{\alpha-2} i^{-\alpha} \geqslant \sum_{S=1}^{K} S^{\alpha-2} \geqslant \frac{K^{\alpha-1}}{\alpha-1}$$

$$\sum_{i=1}^{n} \sum_{S=i}^{K} S^{\alpha-2} i^{-\alpha} \leqslant \sum_{i=1}^{n} \frac{(K+1)^{\alpha-1} - i^{-\alpha}}{\alpha-1} + i^{-2} \leqslant \frac{(K+1)^{\alpha-1}}{\alpha-1} \sum_{i=1}^{\infty} i^{-\alpha} + \sum_{i=1}^{\infty} i^{-2}$$

This means that the first sum in Eq. (C4) shows an asymptotic behaviour as $K^{\alpha-1}$ for $\alpha > 1$.

For $\alpha < 1$ we find

$$\sum_{i=1}^{n} \sum_{S=i}^{K} S^{\alpha-2} i^{-\alpha} \geqslant \sum_{i=1}^{n} \frac{i^{\alpha-1}}{1-\alpha} i^{-\alpha} \geqslant \frac{1}{1-\alpha} \int_1^n \frac{di}{i} = \frac{\ln n}{1-\alpha}$$

$$\sum_{i=1}^{n} \sum_{S=i}^{K} S^{\alpha-2} i^{-\alpha} \leqslant \sum_{i=1}^{n} \frac{i^{-1}}{1-\alpha} + i^{-2} \leqslant \frac{1}{1+\alpha}\left(1 + \int_1^n \frac{di}{i}\right) + \sum_{i=1}^{\infty} i^{-2} = \frac{1 + \ln n}{1-\alpha} + \frac{\pi^2}{6}$$

Thus, the first sum in Eq. (C4) shows an asymptotic dependence on K as ln n.

In the second sum in Eq. (C4), we have $l > K$. For the model in Eqs. (19a), (11a) is $\lambda_l = 0$ for $l \geqslant K$. In this case the second sum vanished. For the model in Eqs. (10), (11) it is $\lambda_l/\mu_l \leqslant 1$ and $\mu_l \geqslant \frac{1}{2}\sigma^2 l^2$. Thus the second sum is

$$\left(\sum_{i=1}^{n} \prod_{l=i}^{K} \frac{\lambda_l}{\mu_l}\right) \sum_{S=K+1}^{\infty} \frac{1}{\mu_S} \prod_{l=K+1}^{S-1} \frac{\lambda_l}{\mu_l} \leqslant c \sum_{i=1}^{n} \left(\frac{K}{i}\right)^{\alpha} \sum_{S=K+1}^{\infty} \frac{2}{\sigma^2 S^2}$$

$$\leqslant c \frac{2}{\sigma^2}\left[\left(\frac{K}{1}\right)^{\alpha} + \int_1^n \left(\frac{K}{i}\right)^{\alpha} di\right] \int_K^{\infty} \frac{1}{S^2}\, ds$$

$$= c \frac{2}{\sigma^2}\left(K^{\alpha} + \frac{K^{\alpha} - \left(\frac{K}{n}\right)^{\alpha}}{\alpha - 1}\right) \frac{1}{K} \leqslant c \frac{2}{\sigma^2}\left(1 + \frac{1}{|\alpha - 1|}\right) K^{\alpha-1}.$$

For $\alpha < 1$, this sum can be dropped compared to the first sum which is proportional $\ln K$ for $n = K$. For $\alpha > 1$ this sum shows the same asymptotic behaviour as the first sum, which is proportional to $K^{\alpha-1}$.

Appendix D

The complete numerical solution of Eq. (3) can be achieved by solving the corresponding eigenvalue problem (Wissel 1984, 1989a). In this case the solution of Eq. (3) has the form (Wissel 1989a)

$$P_n(t) = \sum_\alpha v_n^{(\alpha)} e^{\omega_\alpha t} C^\alpha, \tag{D1}$$

where ω_α are eigenvalues and $v_n^{(\alpha)}$ are eigenvectors of the right side of Eq. (3). The coefficients C_α are determined by the initial condition. For our model in Eqs. (10), (11) and (10a), (11a), all eigenvalues are real and negative. The absolute of the first eigenvalue is found to be much smaller than the absolute of all other eigenvalues.

$$|\omega_1| \leqslant |\omega_\alpha| \quad \text{for } \alpha \geqslant 2. \tag{D2}$$

Therefore the solution (Eq. (D1)] tends to

$$P_n(t) \to P_n^* = v_n^{(1)} e^{\omega_1 t} C_1,$$

a very slowly changing probability distribution, the so-called quasi-stationary distribution P_n^*.

Appendix E

Let the probability that a newborn individual is a female be 1/2. Then the probability of finding j females in between i individuals is given by the binomial distribution

$$W_i(j) = \frac{i!}{(i-j)!j!}\left(\frac{1}{2}\right)^i. \tag{E1}$$

If the probability $A_{k,ij}$ that j females in a population of i individuals have k propagules in the next year is given, we find instead of Eq. (21)

$$P_k(t+1) = \sum_i \sum_{j=0}^{i} A_{k,ij} W_i(j) P_i(t). \tag{E2}$$

The transition matrix $A_{k,ij}$ may be given by a Poisson distribution like Eq. (22) with a mean value as found in Eq. (25). Defining

$$\sum_{j=0}^{i} A_{k,ij} W_i(j) = A_{ki}, \tag{E3}$$

(E2) is reduced to an equation of the form of Eq. (21).

For the explicit model we have to combine different probabilities. The probability of finding l eggs is given by Eq. (27):

$$W_{E,r}(l) = c\exp\left[-\frac{1}{2}\left(\frac{1-l/E}{s}\right)^2\right], \tag{E4}$$

where the constant c is determined by the normalization and the mean value E is given by

$$E = 2Rr, \tag{E5}$$

where r is the number of fertilized females. If $u(l)$ is the probability of survival of a single egg to the adult stage, the probability $W_l(k)$ that k of l individuals will survive is given by the binomial distribution

$$W_l(k) = \frac{l!}{(l-k)!k!} u(l)^k[1-u(l)]^{l-k}. \tag{E6}$$

Equation (29) shows the probability $W(i,j)$ that a single female will be fertilized if j females and $(i-j)$ males are present. Thus the probability that r of j females become fertilized is

$$W_{i,j}(r) = \frac{j!}{(j-r)!r!} W(i,j)^r[1-W(i,j)]^{j-r}. \tag{E7}$$

Now we can combine Eq. (E3) to Eq. (E7) and find:

$$A_{k,i} = \sum_{l=0}^{\infty}\sum_{r=0}^{j}\sum_{j=0}^{i} W_l(k)W_{E,r}(l)W_{i,j}(r)W_i(j). \tag{E8}$$

The mean first passage time (life time) T_n from n to the absorbing state 0 can be determined by the following formula:

$$T_n = \sum_{r\geq 0} T_r A_{rn} + 1 \quad \text{for } n \neq 0. \tag{E9}$$

This means that one needs at least one time step if $n \neq 0$. If one goes to the state r (probability A_{rn}), one needs in addition the mean passage time T_r from this state to zero. Remember that $T_0 = 0$. The system of linear equations (E9) for the T_n can be solved numerically.

Let R_{nm} be the probability of reaching n before zero is reached if one starts at m. The system of linear equations for R_{nm} is

$$R_{nm} = \sum_{r\neq n} R_{nr}A_{rn} + A_{nm}. \tag{E10}$$

This means that n is reached directly from m (probability A_{nm}) or another state r is reached first. We use the notation $W_n = R_{n1}$ and $R_m = R_{Km}$.

As in the model with overlapping generations, we expect that the first eigenvalue ω_1 of the transition matrix A_{k1} shows an absolute value which is much smaller than the absolute of the other eigenvalues. As in Appendix D, we can argue that after a short initial phase the solution $P_n(t)$ of Eq. (21) takes the form:

$$P_n(t) = C_n e^{t \ln \omega_1}.$$

The probability that the population is still alive is

$$P_a(t) = \sum_{n>0} P_n(t) = c e^{t \ln \omega_1}.$$

This exponential decay gives a mean life time

$$T_K = 1/\ln \omega_1 := 1/\omega.$$

The probability that the population is extinct is

$$P_0(t) = 1 - P_a(t) = 1 - c e^{\omega t}.$$

References

Gabriel W, Bürger R (1992) Survival of small populations under demographic stochasticity. Theor Popul Biol 63: 44–71

Goel NS, Richter-Dyn N (1974) Stochastic models in biology. Academic Press, New York

Goodman D (1987a) The demographic of chance extinction. In: Soulé ME (ed) Viable populations for conservation. Cambridge University Press, Cambridge

Goodman D (1987b) Considerations of stochastic demography in the design and management of biological reserves. Nat Res Model 1: 205–234

Jacobs J (1984) Cooperation, optimal density and low density threshold: yet another modification of the logistic model. Oecologia 64: 389–395

Shaffer ML (1981) Minimum population sizes for species conservation. BioScience 31: 131–134

Soulé ME (1987) Viable populations for conservation. Cambridge University Press, Cambridge

Turelli M (1977) Random environments and stochastic calculus. Theor Popul Biol 12: 140–178

Wissel C (1984) Solution of the master equation of a bistable reaction system. Physica 128A: 150–163

Wissel C (1989a) Metastability, a consequence of stochastics in multiple stable population dynamics. Theor Popul Biol 36: 296–310

Wissel C (1989b) Theoretische Ökologie. Eine Einführung. Springer, Berlin Heidelberg New York

Wissel C (1992) Aims and limits of ecological modelling exemplified by the island theory. Ecol Model 63: 1–12

Wissel C, Schmitt T (1987) How to avoid extinction of populations optimally exploited. Math Biosci 84: 127–138

Wissel C, Stöcker S (1991) Extinction of populations by random influences. Theor Popul Biol 39: 315–328

7 "Very Small" Bird Populations in Amazonia

J. Haffer

7.1 Introduction

In view of the vast, limitless expanse of the Amazonian rainforest region, the title of this chapter may appear as a contradiction in itself. However, there are a number of Amazonian habitat types with a conspicuously patchy distribution, where several ecological specialist bird species, as well as other animals, form fairly small and widely separated populations. On the other hand, most rainforest birds of the Amazon basin indeed form large and more or less continuously distributed populations which occupy large areas, although population densities are very low. This has been known in general terms for a long time, but quantitative data of population densities have been obtained only in recent years (Terborgh 1986; Bierregaard and Lovejoy 1989, Terborgh et al. 1990). Among a total of 215 and 145 forest birds studied at two rainforest sites in upper and central Amazonia, respectively, most species were rare, only few species were common and conspicuous. In the upper Amazonian bird community, median abundance was 2.5 pairs per km^2, 84 species had population densities of $\leqslant 1$ pair per km^2 (Terborgh et al. 1990).

Widespread Amazonian forest birds are either ecological generalists or specialists, the latter occupying particular widespread habitats like forest undergrowth, canopy, vine tangles, *Heliconia* thickets, treefall gaps, successional vegetation zones along river courses and others.

7.2 "Very Small" Bird Populations in Amazonia

"Very small" bird populations, as here understood, consist of a comparatively restricted number of pairs, the populations occurring at widely scattered areas. The continued existence of such small populations poses problems regarding recruitment, dispersal behaviour and other demographic aspects of these species. Among Amazonian birds with ecologically restricted distributions which form comparatively small localized populations are species inhabiting patchy habitats like (1) non-forest "islands" of campina vegetation in the rainforest, (2) palm swamps and (3) bamboo thickets in the rainforest, (4) sandbar islands within the Amazon River and its large tributaries as well as (5) isolated low mountains in peripheral regions of Amazonia.

Several aspects of the occurrence of these ecologically specialized bird species of the Amazon basin are discussed in the following section, although no quantitative data on population size at particular habitat patches or on the dispersal behaviour of particular species are available so far. Among monophagous or oligophagous insects, flightless grasshoppers occur in comparatively small populations in lightgaps (treefall sites or other forest openings). Their dispersal to new habitats is a major unsolved problem (Rowell 1983). No facultative winged forms are known. Presumably, individuals walk through the forest until they find a new spot. This occurs when succession in the light gap eliminates their food plant within a few years and makes it essential to find a new habitat.

7.2.1 Birds on Campina "Islands" in the Rainforest

Throughout Amazonia, patches of fairly low, scleromorphic vegetation (low open forest to savannas) occur on deep sandy soils. Such insular campinas vary in size from less than 1 ha to over several 100 ha (and up to 5 million ha). They are particularly common in the upper Rio Negro region and south of the Amazon River, in the Cachimbo region between the upper Rio Tapajós and Rio Xingú, as well as in the Humaitá region west of the upper Rio Madeira. The campina vegetation is characterized by fairly low diversity and high endemicity (Anderson 1981). Oren (1981, 1982) has studied the avifauna of numerous isolated campinas in Amazonia. Below is a summary of some of the results of his work.

The avifauna of the campinas is not very rich, but there are several endemic species and subspecies. Among the 111 species of birds, there are tyrant-flycatchers (25 species), antbirds (13), finches (8), tanagers (8), woodcreepers (7), manakins (6) and others. Plant species' richness and habitat diversity of campina vegetation increase only little with increasing patch size of the campinas. Several bird species are highly dispersive; taking this and other correlated factors into consideration, Oren (1981) estimated that about one-half of the observed increase in species number with increasing campina area can be attributed to area alone. The proportion of ecological specialists increases with campina area and the proportion of generalists decreases. The latter dominate small campinas, since they can also survive in the surrounding forest vegetation. The avifauna of small campinas (less than 3 ha) consists entirely of habitat generalists of very wide neotropical distribution. Campinas 5 to 100 ha in size are inhabited by generalists and some specialists (species strictly limited to campina vegetation). The bird fauna of campinas greater than 100 ha includes, besides generalists, many habitat specialists like the paraqueet *Aratinga aurea*, the piculet *Picumnus varzeae* and the hummingbird *Phaethornis nattereri*. These habitat specialists tend to have restricted geographical ranges (Oren 1982).

$A_{0.5}$ is the patch area at which a species reaches 50% frequency, i.e. the species is present on half of the campina patches of that size; J is the incidence of occurrence of a given species in a particular size class (<10 ha, 10–100 ha,

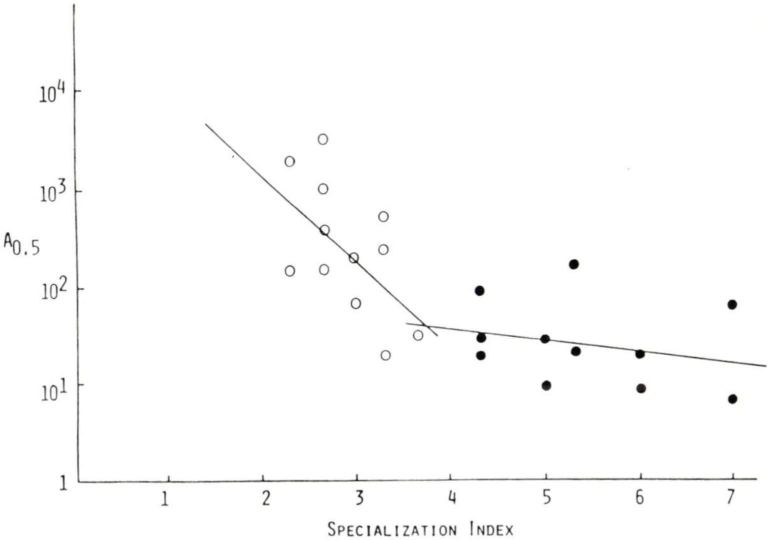

Fig. 7.1. Relationship between specialization index and $A_{0.5}$(ha) for specialists (*open circles*, index values under 4) and generalists (*solid circles*, index values over 4). (From Oren 1981)

101–1000 ha, 5000–50 000 ha, >1 000 000 ha). Due to the ecological requirements and the dispersal capabilities of the different species, their population size on campinas smaller than the species-specific value of $A_{0.5}$ apparently is too small for their continued existence. Estimates of population size of bird species on campina islands of different areal extent are not available.

The relationship between degree of ecological specialization of campina birds (based on diet and habitat preferences, Oren 1981) and the minimum area required for a given species to survive in campinas of different sizes is shown in Fig. 7.1, where $A_{0.5}$ is plotted versus specialization index. Specialists (with low index values) have higher minimum area requirements as measured by $A_{0.5}$ than do generalists ($t = -4.354$, significant at $p = 0.0003$). Because habitat generalists can exist in surrounding habitats, the campina size in which they first reach 50% occurrence ($A_{0.5}$) is smaller than for comparable habitat specialists, which can exist only in campina vegetation. The latter species therefore require larger campina islands for survival.

Campina habitat specialists which occur in the smallest campinas (lowest $A_{0.5}$ values) take both plant and animal food (e.g. *Xenopipo atronitens*, *Neopelma chrysocephalum*, *Heterocercus flavivertex* and *Elaenia ruficeps*); specialists which take only animal food or only plant food have higher $A_{0.5}$ values. There is no endemic campina bird of prey; even the largest campina area is apparently too small an area to support a habitat specialist raptor in campinas. The raptors which do occur in campiñas are habitat generalists.

7.2.2 Birds of Palm Swamps in the Rainforest

The point-tailed palmcreeper (*Berlepschia rikeri*) is a medium-sized member of the species-rich suboscine family Furnariidae and an endemic Amazonian species. Because of its many morphological and ecological peculiarities, this bird is placed in the monotypic genus *Berlepschia*. As a strict habitat specialist, it occurs exclusively in association with *Mauritia* palm groves (Olalla 1935; Vaurie 1980). These tall palms, which grow in swampy areas of the rainforest as well as along rivers and other bodies of water, have large crowns. Such palm groves are the home of pairs or family groups of *Berlepschia*. This bird is said to alight only on these palms, never on other trees. Its food is predominantly insects gleaned from the palm leaves. *Berlepschia* is known from several widely scattered localities throughout the Amazonian-Guianan forest region. Even though the species probably occurs also in regions between known collecting sites, it is certainly of very local patchy occurrence, one or a few pairs inhabiting a local palm grove in the forest which is widely separated from the next groups of pairs in another palm grove. Near the western margin of the species' range in southeastern Peru and eastern Bolivia, Parker et al. (1991) found two pairs along the edge of a lagune and several pairs in *Mauritia*-dotted savannas. They also observed two pairs in a small *Mauritia* swamp surrounded by hill forest and many kilometres from the nearest suitable habitat. In some portions of central Amazonia, e.g. near Manaus, *Mauritia* palm groves are found more frequently along forest creeks and small rivers, thus supporting somewhat larger populations of *Berlepschia* (D.F. Stotz, pers. comm.).

Nothing is known about the dispersal behaviour of this bird, which enables it to search for new habitats and to maintain very small populations at widely scattered localities.

7.2.3 Birds of Bamboo Thickets in the Rainforest

The rainforests of southwestern Amazonia in southeastern Peru and adjacent western Brazil are especially rich in bamboo, which occurs along the banks of major rivers and on the tops of low hills between small stream valleys. Regionally, the bamboo habitat is patchy, and habitat specialists among birds restricted to bamboo thickets occur in small populations separated by long distances from other populations. Examples of bamboo specialists among birds include the antbirds *Cymbilaimus sanctaemariae*, *Cercomacra manu* and *Drymophila devillei*, as well as the flycatchers *Poecilotriccus albifacies*, *Hemitriccus flammulatus*, *Lophotriccus eulophotes* and *Ramphotrigon megacephala*. A fairly large number of additional, ecologically more widespread species in these regions inhabit bamboo stands, although they are not restricted to them (Parker 1982; Pierpont and Fitzpatrick 1983; Fitzpatrick and Willard 1990; Parker et al. 1993). The spotty distribution of bamboo specialist birds in Amazonia certainly reflects the patchiness of their preferred bamboo habitat in this region. So far, however, no quantitative details are known on the size of bamboo habitat patches in Amazonia or on the distance between individual habitat patches. Details on the population size of bamboo specialists in individual habitat patches and on their dispersal behaviour also remain unknown.

7.2.4 Birds on Sandbar River Islands

About 100 bird species (15% of the non-aquatic avifauna of the Amazon basin) are restricted to habitats created by rivers, a situation unparalleled by any other avifauna of tropical or temperate river systems of the world (Remsen and Parker 1983). Among these birds, 15–20 species inhabit exclusively ephemeral sandbar islands in the Amazon River and in the wide lower portions of several large tributary streams. These species often form very small populations relative to species of more widely distributed habitats. The river islands are covered by early successional habitats and undergo continuous erosion at the upstream end and grow through deposition of sand and silt at the downstream end, thus slowly "migrating" with the river current. Young sandbar river islands are covered with the composite shrub *Tessaria integrifolia* and the willow *Salix humboldtiana*. Older and larger islands are dominated by extensive stands of tall and short Cecropia forest. During the high-water season when the water level rises 5–10 m, the *Tessaria* and *Salix* stands, as well as the understorey of the *Cecropia* forest, are usually under water.

Among a total of 231 species recorded on middle-aged and old river islands (over 3 years old) in northeastern Peru, most species also occurred on the mainland [33% in second growth, 25% in seasonally flooded forest (varzea) and 40% in water-dependent (riverine) habitats like river, stream and lake margins, sandbars and marsh; Rosenberg 1990]. However, 18 species are obligate river-island birds which, in this region, have never been encountered on the mainland. Some of these species also inhabit dry patchy habitats, second-growth, river-edge habitats or sand ridges in other portions of Amazonia (where several additional species are on the list of obligate river-island birds).

Most of the obligate river-island species are representatives of genera typically found in open habitats (the hummingbirds *Leucippus*, the furnariids *Furnarius* and *Certhiaxis*, the flycatchers *Elaenia*, *Stigmatura*, *Serpophaga*) or a mixture of forest and open habitats (*Synallaxis*, *Cranioleuca*, *Thamnophilus*, *Cnemotriccus*, *Conirostrum*). Only *Myrmotherula assimilis* and *Myrmoborus lugubris* are representatives of mostly forest genera. *Elaenia pelzelni* and two species of *Conirostrum* (*C. bicolor*, *C. margaritae*) are restricted to the canopy of tall *Cecropia* forest, several other obligate river-island species are restricted largely to *Tessaria* scrub on sandbars (*Synallaxis propinqua*, *Stigmatura napensis*) or to taller willows (*Salix*) associated with this scrub habitat (*Serpophaga hypoleuca*). The antbird *Myrmoborus lugubris* is restricted to the understorey of tall *Cecropia* forest. Other obligate island species are ecological generalists occurring in more than one microhabitat, but foraging primarily in the understorey of the *Cecropia* forest within 3 m of the ground, e.g. the antbirds *Myrmotherula assimilis* and *Myrmochanes hemileucus*.

The fact that the river islands are under water during flood season requires a high degree of mobility and dispersal for the river-island species to take advantage of these ephemeral island habitats as they become available during the low-water season, a higher degree of dispersal capability than in most tropical understorey birds. However, almost nothing is known about the dispersal or the temporal and spatial movements of obligate island species in response to the seasonal inundation of their habitat. During the high-water season when the island floor and the understorey of the *Cecropia* forest are flooded and only the tree tops are above water level, field observations suggest that some species leave the islands, whereas canopy species and certain understorey birds may stay, the latter (e.g. *Furnarius minor*) undergoing vertical shifts depending on the height of the water level above the ground. Similarly, the successional growth of the island vegetation forces some island species to search for new habitats when, for example, *Tessaria* scrub vegetation gives way to *Cecropia*-dominated forest. Many birds can disperse and wander from island to island in search of their preferred habitat. Even a poor-flying terrestrial tinamou (*Crypturellus undulatus*) has been observed to fly across open water from the mainland to an island, and the antbirds *Myrmoborus lugubris*, *Myrmochanes hemileucus* and *Thamnophilus cryptoleucus* cross rivers up to 2 km wide to reach river islands during the low-water season (Remsen and Parker 1983).

The existence of stable populations of several of these obligate island species depends on the continued formation and presence of new sandbar islands. It appears that the numerous small populations of obligate island species along the Amazon river system seasonally form larger populations through the birds' dispersal behaviour during the flood season.

7.2.5 Birds on Isolated Low Mountains

A number of bird species are restricted to small and isolated mountain ridges around the Amazonian perimeter (Fitzpatrick 1976, 1980; Fitzpatrick and O'Neill 1979). Quantitative details on population size remain unknown. Fitzpatrick (1976) hypothesized that, as successful species dispersed through a continuous Amazonian forest, competitively inferior forms were forced to occupy progressively smaller ranges, and surviving populations were increasingly fragmented. In the final stages, a species may exist only in relatively competitor-free regions, such as on the isolated peripheral mountain ridges to which they are restricted today. Examples among pygmy flycatchers (Tyrannidae) are *Idioptilon rufigulare* and *I. cinnamomeipectus*, rare montane relict species on isolated eastern Andean foothill ridges in Peru. Other representative *Idioptilon* species occur in extremely restricted and widely disjunct areas of eastern Brazil.

Another example is the sharpbill (*Oxyruncus cristatus*), a taxonomically isolated and monotypic genus (of the flycatchers or cotingids), which occurs in southeastern Brazil and in widely scattered and relatively small populations along mountain ridges mostly around the Amazonian lowlands (Chapman 1939), including low ranges (300–800 m) of Amapá and Carajás (Silva 1993). In recent years, *Oxyruncus* has also been found in upland forests (550 m elevation) on isolated ridges in Alagôas, northeastern Brazil (Teixeira et al. 1986) and at middle elevations along the eastern slope of the Andes of Peru and Bolivia, where it has a patchy distribution, mainly in stunted forests on rather sandy soils near the crest of outlying ridges (Parker et al. 1991).

The male plumage of the white bellbird (*Procnias alba*) is, as the name suggests, snow-white; the female is cryptically coloured. This medium-sized cotingid species inhabits lower montane and montane forests in northeastern South America (southeastern Venezuela and Guianas). A widely isolated and fairly small population (by Amazonian standards) of *P. alba* far to the south of the lower Amazon River was discovered recently in the Serra dos Carajás (Roth et al. 1984; Oren and Novaes 1985). This area is about 1000 km south of the nearest known conspecific populations and separated by ecologically unsuitable lowland rainforest in the wide Amazon Valley. The Serra dos Carajás, a region of about 18 000 km^2 between the Araguaia/Tocantins and Xingú Rivers, is centered at approximately 6°00'S, 50°30'W. Several low plateau reach 700–800 m elevation. White bellbirds occur in the forests at altitudes of 500 to 750 m, where they are locally common, but with patchy distribution. This isolated population is subspecifically differentiated from the northern main population of this species.

7.2.6 Rare and Localized Amazonian Forest Birds

There are a number of very rare Amazonian bird species known only from the type locality, although their ranges are probably at least somewhat more extensive than the available records suggest. Examples are the small flycatcher *Todirostrum senex* (lower Rio Madeira), the antbirds *Thamnophilus praecox* (southeastern Ecuador), *Pithys castanea* (northeastern Peru) and *Clytoctantes atrogularis* (Rondônia, western Brazil), the manakin *Pipra vilasboasi* (Rio Cururú, Serra do Cachimbo), the spinetail *Poecilurus kollari* in the savanna country of the upper Rio Branco region (northern Bràzil). Nothing is known about the population size of these geographically very restricted species. The small flycatcher *Hemitriccus aenigma*, until recently reported only from the lower Rio Tapajós, actually is widely distributed in south-central Amazonia (Bates et al. 1992).

Several additional enigmatic forest birds, known only from the holotype, may not be valid, possibly representing interspecific hybrids or aberrant plumage types of known and widespread species. Among such rare antbirds (Formicariidae) is *Phlegopsis barringeri* from southeastern Colombia, which Willis (1979) and Graves (1992) suggested is based on an interspecific hybrid. Schulenberg and Stotz (1991) showed that another antbird known from the

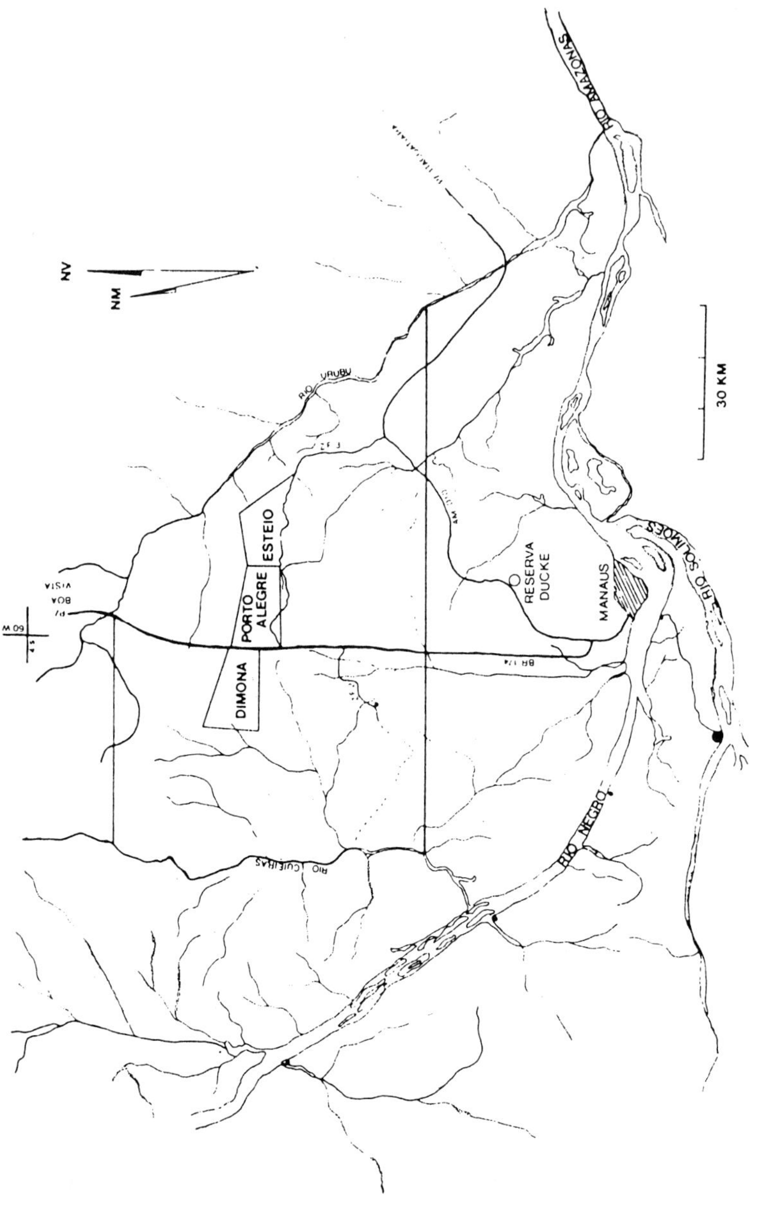

Fig. 7.2. Manaus in central Amazonia and three cattle ranches (Dimona, Porto Alegre, and Esteio), where the Biological Dynamics of Forest Fragments Project reserves are located. (From Bierregaard and Lovejoy 1988)

unique type *Myrmeciza "stictothorax"* is a name based on a plumage abnormality of a male of the widespread species *M. atrothorax*.

7.2.7 The Biological Dynamics of Forest Fragments Project

This project was initiated in 1979 under the designation Minimum Critical Size of Ecosystems, a joint project of Brazil (INPA, Manaus) and the United States (World Wildlife Fund, WWF, Washington) to study the effects of fragmentation on rainforest ecosystems. In 1989, the project moved to the Smithsonian Institute and is being funded by several sponsoring agencies including the WWF. The life of the project at its initiation was planned to be around 20 years to make possible the long-term monitoring of fragmentation effects. A series of forest reserves of different size (1, 100, 1000 ha, etc.) are being studied prior to and after isolation in central Amazonia about 80 km north of Manaus (Fig. 7.2). The project's name was changed recently to Biological Dynamics of Forest Fragments Project (BDFF). It focusses on how scale (size of forest fragments) and rates of species loss relate to subsequent species composition (Bierregaard 1986; Bierregaard and Lovejoy 1988, 1989; Bierregaard and Hutchings 1990; Lovejoy and Bierregaard 1990; Bierregaard et al. 1992).

The microclimate is being monitored in the isolated reserves, and detailed botanical studies of certain plant families, particularly palms and trees, are being carried out to determine, for example, the ecological parameters which influence the local occurrence of these plant species and the minimum population sizes and reserve sizes necessary for their continued survival. A broad range of animal taxa are being studied under this project, e.g. understorey birds, ant-following birds, mixed-species flocks of forest birds, herpetofauna, mammals, butterflies, ants and other groups of insects. Questions addressed are, among others: how does habitat fragmentation affect diversity, distribution and abundance of animal species? How do activity patterns of resident animals change upon isolation? What species and ecological guilds are most susceptible to fragmentation effects and why are they sensitive? How do certain species persist in small forest fragments?

Regarding plants, the data from isolated reserves show a high mortality of adult trees along the reserve edges and an increase in standing dead trees in the centre of at least one 10-ha isolate. Other isolation effects are seen in a 100-ha isolated forest fragment. In the case of understorey bird communities, the most immediate and dramatic effect of isolating 1-ha and 10-ha forest reserves was the observed greater foraging activity of birds in a reserve whose area is smaller than the birds' normal home ranges (increase of capture rates) and an influx of displaced individuals fleeing the felled forest (crowding effect leading to a decrease in recapture rates). Individuals of the obligate ant-following species with home ranges of 1–2 km in diameters, therefore, did not persist in 10-ha forest fragments (ca. 350 m across); see Figs. 7.3 and 7.4.

Some mid-storey forest birds show behavioural shifts in foraging height after forest isolation, thereby increasing their vertical foraging range to compensate

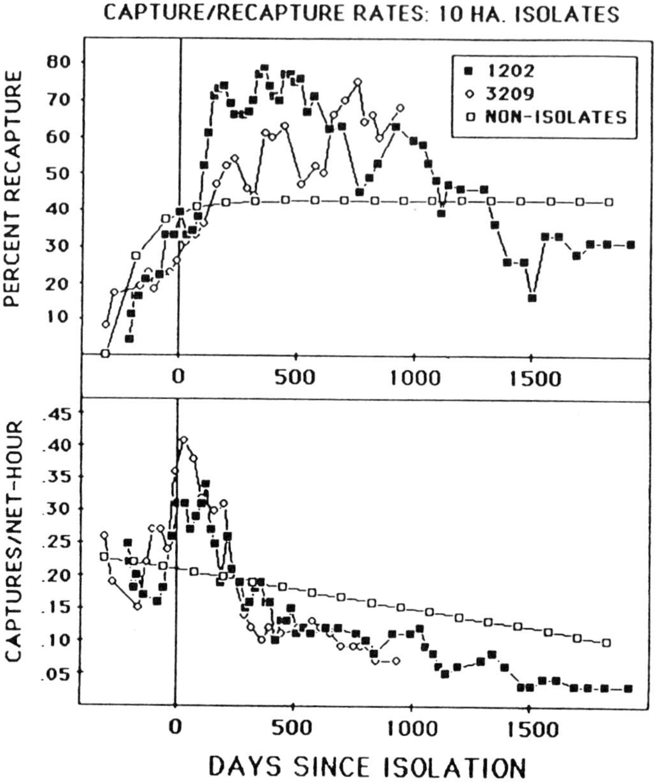

Fig. 7.3. Recapture percentages and capture rates of understorey forest birds in two 10-ha reserves (nos. 1202 and 3209) before and after isolation. Time of isolation (*0 days since isolation*) is set at the approximate midpoint between beginning and finishing of felling around the reserves, so some response to felling is expected before t = 0. Expected values calculated from data from net-lines in continuous forest are plotted for comparison. To smooth the curves, data for isolated reserves are three-point running means. (From Bierregaard and Lovejoy 1988)

for the forest reduction in horizontal space. Ground-foraging birds cannot adjust to forest isolation and reduction in this manner. The complex mixed-species insectivorous flocks typical of Amazonian forests deteriorated within 2 years of isolation of 1-ha and 10-ha forest fragments. Movement of forest birds to and from a reserve is severely limited if a reserve is isolated from continuous forest by less than 100 m of cleared area. Therefore, forest corridors linking isolated reserves to continuous forest increase the conservation value or effectiveness of an isolated forest reserve. Species are lost faster from small reserves than from larger ones in a predictable pattern ("ecosystem decay"). The lack of invasion of any second-growth bird species into the forest from the ecotone results in an abnormally low level of avian activity along forest edges. This pattern is reversed in forests of the temperate zone.

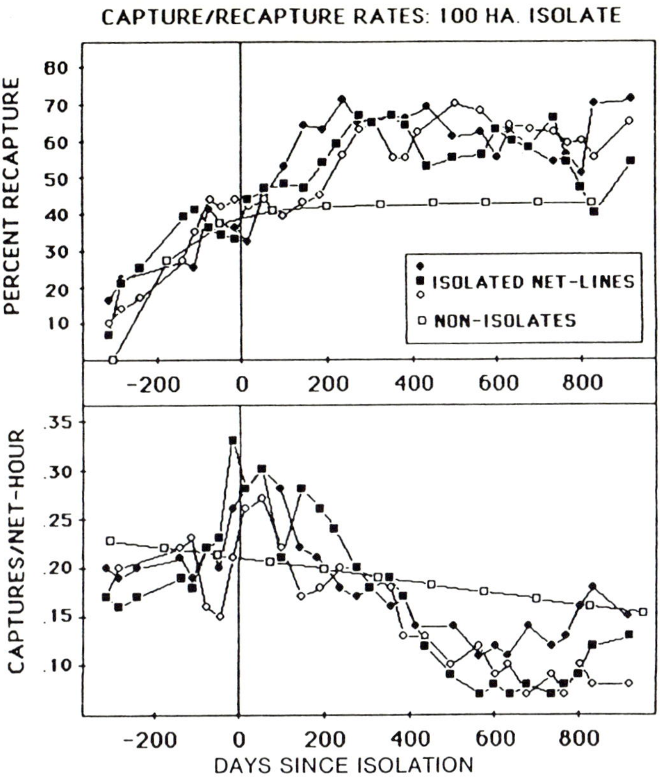

Fig. 7.4. Recapture percentages and capture rates of three net-lines in a 100-ha reserve before and after isolation; see also legend to Fig. 7.2. (From Bierregaard and Lovejoy 1988)

The results of these long-term studies are highly relevant for conservation planning, as the Amazonian landscapes including terra firme forest at a distance from the main rivers are being increasingly "developed" for human use. Of similar importance for conservation planning will be the knowledge of faunas inhabiting the campina islands of the Amazonian lowlands. River-created habitats, especially sandbar islands of the Amazon River and its major tributaries, also need protection, since some island species are endangered by higher flooding resulting from deforestation in the Andes.

7.3. Summary

The majority of Amazonian forest birds form large populations which occupy huge areas, although population densities are very low. On the other hand, there are a number of ecological specialists among birds and other animals which occupy patchy habitats in Amazonia where they form rather small isolated populations. In islands of non-forest campina vegetation, ecological specialists have higher minimum area requirements than do generalists because the latter can exist also in surrounding habitats. Some bird species are narrowly adapted to palm swamps (*Mauritia* sp.) or bamboo thickets and therefore have patchy distributions in Amazonia. There are about 15–20 bird species which, in Amazonia, inhabit exclusively successional habitats on ephemeral sandbar islands in the largest rivers, where they form fairly small populations during the breeding season. During the high-water season when the island floor and the understorey of the vegetation are flooded and only the treetops are above water level, some obligate sandbar island species leave the islands, whereas others may stay. Some understorey birds undergo vertical shifts, depending on the height of the water level above the ground. To take advantage of the ephemeral island habitats as they become available during the low-water season requires a higher degree of dispersal capability than most tropical understorey birds possess.

A 20-year joint Brazilian and United States project in central Amazonia near Manaus is designed to monitor the effects of forest fragmentation on rainforest ecosystems. The project focusses on how scale (size of forest fragments) and rates of species' loss relate to subsequent species' composition. A broad range of plant and animal taxa are being studied (birds, herpetofauna, mammals, various groups of insects). The results of these long-term studies and the data on faunas inhabiting patchy Amazonian habitats of campina vegetation, bamboo thickets and sandbar islands are relevant for conservation planning in Amazonia.

Acknowledgements. I gratefully acknowledge discussions with D.C. Oren (Belém, Pará) and D.F. Stotz (São Paulo) on the occurrence of small populations of birds in Amazonia. R.O. Bierregaard (Washington) kindly provided me with reprints of publications and other material on the results of the joint INPA-Smithsonian project in central Amazonia north of Manaus. K. Riede (Freiburg, Germany) sent me copies of several relevant entomological publications.

References

Anderson AB (1981) White sand vegetation of Brazilian Amazonia. Biotropica 13: 199–213

Bates JM, Parker III TA, Capparella AP, Davis TJ (1992) Observations on the campo, cerrado and forest avifaunas of eastern Dpto. Santa Cruz, Bolivia, including 21 species new to the country. Bull Br Ornithol Club 112: 86–98

Bierregaard RO (1986) Changes in bird communities in virgin forest and isolated Amazonian forest fragments. Ibis 128: 166–167

Bierregaard RO, Lovejoy TE (1988) Birds in Amazonian forest fragments: effects of insularization. Acta XIX Congr Int Ornithol 2: 1564–1579

Bierregaard RO, Lovejoy TE (1989) Effects of forest fragmentation on Amazonian understorey bird communities. Acta Amazonica 19: 215–241

Bierregaard RO, Hutchings R (1990) Biological dynamics of forest fragments project; Tenth Annual Report. INPA, Manaus

Bierregaard RO, Lovejoy TE, Kapos V, Santos AA dos, Hutchings R (1992) The Biological dynamics of tropical rainforest fragments. Bio Science 42: 859–866
Chapman FM (1939) The riddle of *Oxyruncus*. Am Mus Novit 1047, 4 pp
Fitzpatrick JW (1976) Systematics and biogeography of the tyrannid genus *Todirostrum* and related genera (Aves). Bull Mus Comp Zool 147: 435–463
Fitzpatrick JW (1980) Some aspects of speciation in South American flycatchers. Acta XVII Congr Int Ornithol 2: 1273–1279
Fitzpatrick JW, O'Neill JO (1979) A new tody-tyrant from northern Peru. Auk 96: 443–447
Fitzpatrick JW, Willard DE (1990) *Cercomacra manu*, a new species of antbird from southwestern Amazonia. Auk 107: 239–245
Graves GR (1992) Diagnosis of a hybrid antbird (*Phlegopsis nigromaculata* x *Phlegopsis erythroptera*) and the rarity of hybridization among suboscines. Proc Biol Soc Washington 105: 834–840
Lovejoy TE, Bierregaard RO (1990) Central Amazonian forests and the minimum critical size of ecosystems project. In: Gentry AH (ed) Four neotropical rainforests. Yale University Press, New Haven, pp 60–74
Olalla AM (1935) *Berlepschia rikeri* y su biologia. Rev Mus Paul 19: 419–423
Oren DC (1981) Zoogeographic analysis of the white sand campina avifauna of Amazonia. PhD Thesis, Harvard University, Cambridge, Massachusetts
Oren DC (1982) Testing the refuge model for South America. A hypothesis to evaluate discrepancies in refuge number across taxa. In: Prance GT (ed) Biological diversification in the tropics. Columbia University Press, New York, pp 601–607
Oren DC, Novaes FC (1985) A new subspecies of white bellbird *Procnias alba* (Hermann) from southeastern Amazonia. Bull Br Ornithol Club 105: 23–25
Parker III TA (1982) Observations of some unusual rainforest and marsh birds in southeastern Peru. Wilson Bull 94: 477–493
Parker III TA, Castillo AU, Gell-Mann M, Rocha OO (1991) Records of new and unusual birds from northern Bolivia. Bull Br Ornithol Club 111: 120–138
Parker III TA, Stotz DF, Fitzpatrick JW (1993) Notes on avian bamboo specialists in southwest Amazonian Brazil. Wilson Bull (in press)
Pierpont N, Fitzpatrick JW (1983) Specific status and behavior of *Cymbilaimus sanctaemariae*, the Bamboo Antshrike, from southwestern Amazonia. Auk 100: 645–652
Remsen JV Jr, Parker III TA (1983) Contribution of river-created habitats to bird species richness in Amazonia. Biotropica 15: 223–231
Rosenberg GH (1990) Habitat specialization and foraging behavior by birds of Amazonian river islands in northeastern Peru. Condor 92: 427–443
Roth P, Oren DC, Novaes FC (1984) The white bellbird (*Procnias alba*) in the Serra dos Carajás, southeastern Pará, Brazil. Condor 86: 343–344
Rowell HF (1983) *Drymophilacris bimaculata*. In: Janzen DH (ed) Costa Rican natural history. University of Chicago Press, Chicago, pp 714–716
Schulenberg TS, Stotz DF (1991) The taxonomic status of *Myrmeciza stictothorax* (Todd). Auk 108: 731–733
Silva JMC da (1993) The Sharpbill in the Serra dos Carajás, Pará, Brazil, with comments on altitudinal migration in the Amazon region. J Field Orn 64: 310–315
Teixeira DM, Nacinovic JB, Tavares MS (1986) Notes on some birds of northeastern Brazil. Bull Br Ornithol, Club 106: 70–74
Terborgh JW (1986) Population densities of Amazonian birds: implications for conservation. Ibis 128: 165–166
Terborgh JW, Robinson SK, Parker III TA, Munn CA, Pierpont N (1990) Structure and organization of an Amazonian forest bird community. Ecol Monogr 60: 213–238
Vaurie C (1980) Taxonomy and geographical distribution of the Furnariidae (Aves, Passeriformes). Bull Am Mus Nat Hist 166: 1–357
Willis EO (1979) Comportamento e ecologia da mãe-de-taoca, *Phlegopsis nigromaculata* (d'Orbigny & Lafresnaye) (Aves, Formicariidae). Rev Bras Biol 39: 117–159

8 Minimum Population of *Clethrionomys glareolus* (Schreber 1780)

G. Bujalska

8.1 Introduction

The term minimum population denotes the smallest population size that ensures persistence: the loss of individuals can be counterbalanced by their gain. However, chance events that affect minimum population (habitat deterioration due to human impact or seasonality of reproduction combined with weather fluctuations, for instance) may increase the risk of its extinction.

It was found for local populations of the bank vole, especially those at the northern border of the species range, that particularly low abundance in spring, i.e. at the beginning of the breeding season, resulted in either a further depression (Hansson and Henttonen 1985) or even their extinction (Bujalska 1985a). This, however, does not seem to affect the existence of this species, because such locally extinct population units are re-established by immigrant individuals, and in 2–3 years the local density may return to the usual range. Local population replacement due to dispersal of individuals (den Boer 1968, 1990) can be considered important for the viability of a population understood as a set of local units (subpopulations).

This chapter is aimed, however, at finding out whether or not, in a confined population of the bank vole that inhabits Crabapple Island, there exist mechanisms enhancing population growth rate in the period of frequently lowest population size, that is in early spring. At the beginning of the breeding season, the size of *this population depends entirely on* the weather and food conditions, and the dying individuals can be replaced neither by immigrants nor newly born voles (Bujalska 1985a; Bujalska and Grüm 1989). Intense reproduction of the overwintered individuals combined with the low mortality rate of their progeny could be interpreted as a means of acceleration of the population recovery.

8.2 Area, Material and Methods

The material was sampled from a population of bank voles (Fig. 8.1) inhabiting Crabapple Island (4 ha in area) in Beldany Lake (Fig. 8.2) (northeastern Poland) in 1966–1970, 1972 and 1975–1990. The study site is placed in the central part of the geographical range of this species (Raczyński 1983). Four forest associations

Fig. 8.1. A bank vole

Fig. 8.2. Crabapple Island from a distance

cover the entire area of the island (Fig. 8.3): in the central, most elevated part there is Tilio-Carpinetum (Traczyk 1962), in two subassociations: Tilio-Carpinetum typicum (Fig. 8.4) and Tilio-Carpinetum stachyetosum (Fig. 8.5) that cover 78% of the entire area. Salici-Franguletum (Malc. 1929) (Fig. 8.6), distributed along the island border, occupies 15% of the area, and the most humid (and the less elevated) part (7%) is occupied by Circaeo-Alnetum (Oberdorfer 1953; Fig. 8.7). A detailed description of the vegetation is given by Traczyk (1965, 1971). Estimates of herb layer biomass (in terms of dry weight)

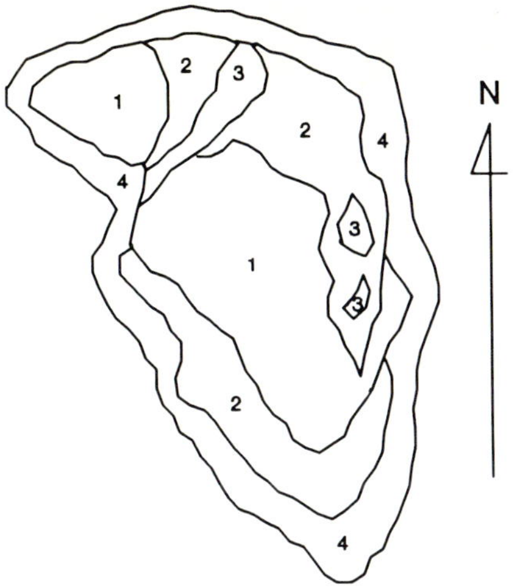

Fig. 8.3. A schematic map of plant associations on Crabapple Island. *1* Tilio-Carpinetum typicum; *2* Tilio-Carpinetum stachyetosum; *3* Circaeo-Alnetum; *4* Salici-Franguletum

Fig. 8.4. The association of Tilio-Carpinetum typicum

are based on 20 samples (each of 0.1 m^2) taken from each plant association during every series of trappings. According to Pucek (1983), the study site provides optimal habitat conditions for bank voles. Bank voles of various distinct groups evenly inhabit all the plant associations. However, individual bank voles stay for a longer time in Tilio-Carpinetum typicum than in Circaeo-Alnetum (Bujalska and Mieszkowska 1984).

The herb layer begins to develop usually at the end of March, and in 1 month it is visibly abundant. At the end of April many species are in bloom, e.g. *Anemone silvestris*, *Hepatica nobilis* and *Corydalis cava*. Their flowers, as well

Fig. 8.5. The association of Tilio-Carpinetum stachyetosum

Fig. 8.6. The association of Salici-Franguletum

as young leaves, are grazed by bank voles, and also transported to the nests to complement the food of the young. Hence, one can suppose that in the latter period bank voles, being mostly plant eaters, have an abundant food supply. The mean value of the herb layer biomass in April (all habitats and years) is 288 (range 32–486 in different years) kg/ha (Bujalska unpubl.).

Reproduction of the island bank voles starts in early April; however, it may begin as early as March 25 or, on the other hand, as late as mid-April. On the average, it can be accepted that reproduction begins on April 1 (Bujalska 1983).

Fig. 8.7. The association of Circaeo-Alnetum

Not all individuals attempt to reproduce simultaneously, and therefore in the last 10 days of April females in oestrus can be observed (they could not be pregnant earlier in the current breeding season) as well as those in the early days of pregnancy. It is also the period of first parturitions (Bujalska et al. 1968; Bujalska 1970, 1983).

Bank voles were trapped live, and newly caught animals were individually marked (toe-clipped). There were 3 traps in each of 159 trap sites arranged in a grid, with 15-m intervals between the trap sites, with the grid covering the entire island area. Each trapping series lasted at least 7 days. Traps were inspected twice daily (at 07.00 and 19.00), and the time lapse between the trapping series was 6 weeks. Each year there were five trapping series in the breeding season, from April to October. In this chapter only the data from the first (April) and second (June) trapping series are analyzed. This is because the most rapid population growth occurs between these series (Bujalska 1985a,b) and, moreover, the lowest seasonal population size as a rule can be found in April.

Population size, i.e. the entire number of the trappable bank voles inhabiting the island, was estimated for each of the trapping series with the so-called common census method, which yields values 5–10% higher than Jolly's (1965) method (Bujalska 1985a).

The number born was calculated from the following formula:

$$NB = (NP\ T\ L)/TP,$$

where NP represents the mean number of pregnant females in a period, and T, L and TP stand for duration of the period (days), litter size (number of individuals) and duration of pregnancy (days), respectively (Bujalska 1975).

NP was obtained from numeric integration of data on the number of pregnant females in consecutive trapping series, accepting that the number of pregnant females at the beginning of the breeding season is equal to 0, and interpolating linearly the number of those pregnant for all the days of the period T. Pregnant females were distinguished on the basis of vaginal smears, changes in body weight, palpation of embryos and subsequent lactation. Each distinct period T ended 21 days before a trapping series, and thus all the young born in a period could be captured in the nearest trapping series. Litter size (varying seasonally around five to a litter) was accepted after Zejda (1966). Duration of pregnancy, equal to 22 days, was accepted after Bujalska and Ryszkowski (1966). All the young born in the period from the beginning of the breeding season to 21 days before the trapping series in June are named cohort K_1.

Survival of the young in the period from their birth to the first capture (named untrappable survival) was assessed from the estimated numbers of those born (K_1) and then the numbers of this cohort caught in June. Either percentage of survivors or instantaneous mortality rate was estimated. The latter is assessed using the time lapse between the so-called mean day of birth (a day until which at least 50% of the young K_1 have been born) and the last day of the trapping series in June (Bujalska 1975).

Survival of the trappable bank voles was estimated for the overwintered animals that were present in the trapping series in April: the percentage of those surviving till June was calculated.

8.3 Results

During 22 study years, the population numbers in April were highly variable (Fig. 8.8). The lowest numbers were found in 1987 (four overwintered individuals), and the highest in 1972 (152 individuals). The mean and standard deviation were 59.5 and 43.5, respectively. The coefficient of fluctuation (Whittaker 1975) calculated for April numbers (obtained for all the study years) was 2.05 (Bujalska 1988). In June, the population numbers varied from 11 individuals (in 1986 and 1987) to 339 (in 1983). The mean and standard deviation were 129.6 and 89.6 individuals, respectively. The coefficient of fluctuation decreased in comparison to that of April and equalled 1.84 (Bujalska 1988).

The above decrease in the coefficient of fluctuation suggests that there is no linear relation between the population size in April and that in June. In fact, the relations between these variables can be described by a second degree parabolic regression (Fig. 8.9). Within the range 4–60 (parabolic regression) of the population sizes in April, there is almost linear growth of the population size in June; however, when the April size exceeds 60 individuals then there seems to be no increasing trend in the population size in June (Fig. 8.9).

More detailed insight into the relations between the population size in April and in June emerges from the regression of the population growth rate (i.e. the ratio of population numbers in June to that in April) on the initial population

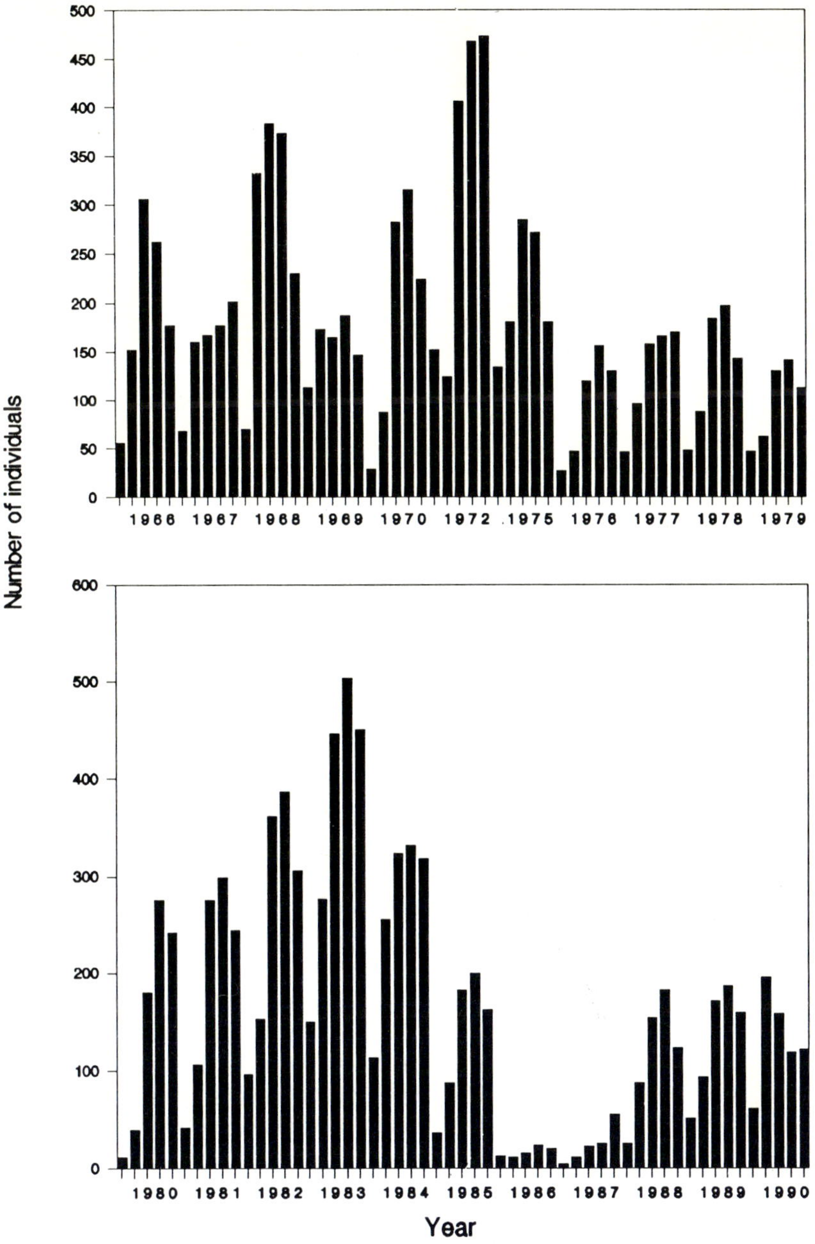

Fig. 8.8. Multiannual dynamics of the bank vole population of Crabapple Island. There are five estimates (in April, June, July, September and October) of population numbers per year

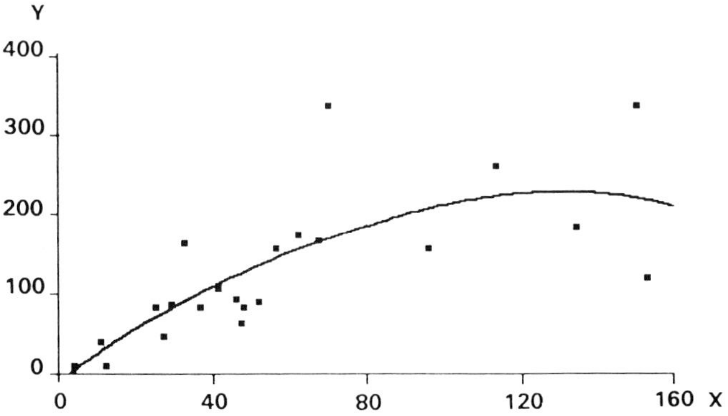

Fig. 8.9. Population size in June as a function of the size in April. Regression formula: $Y = -14.9 + 3.713\ X - 0.0143\ X^2$, $R = 0.722$, $n = 22$, $F = 12.45$, $p = 0.0001$. *Y* Population numbers in June; *X* population numbers in April

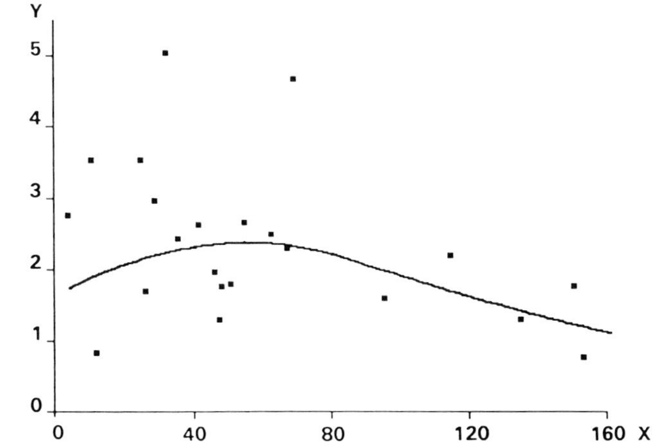

Fig. 8.10. Population growth rate as a function of population size in April. Regression formula: $Y = 1 - [1/(0.43 + 0.00536X - 0.0000489X^2)]$, $R = 0.532$, $n = 22$, $F = 3.76$, $p < 0.03$. *Y* Ratio of population numbers in June compared to that in April; *X* population numbers in April

size (Fig. 8.10). The lowest (<20) and the highest (>90) population size in April results in a lower population growth rate than that obtained for the intermediate initial population size (20–90 individuals). It may indicate that at the lowest population sizes in April, the number of those born will not counterbalance the number that die. This would denote the subminimum initial population size.

Population variables that determine the size in June are evidently as follows: (1) initial (April) population size, (2) number born (K_1), (3) number of overwintered dying from April to June and (4) number of K_1 dying prior to their first capture

Table 8.1. Linear regression of the population growth rate from April to June (Y) on population size in April (X1), number of progeny of overwintered individuals (X2), and on nestling mortality rate (X3)

Predictor variable	Coefficient	Adjusted R-squared	Studen's t	p
Constant	3.233	—	10.64	0.0000
X1	−0.0345	0.0999	−4.24	0.0005
X2	0.0112	0.4643	4.32	0.0004
X3	−52.11	0.6886	−3.83	0.0012

$R = 0.829$, $n = 22$, $F = 16.48$, $p < 0.0001$

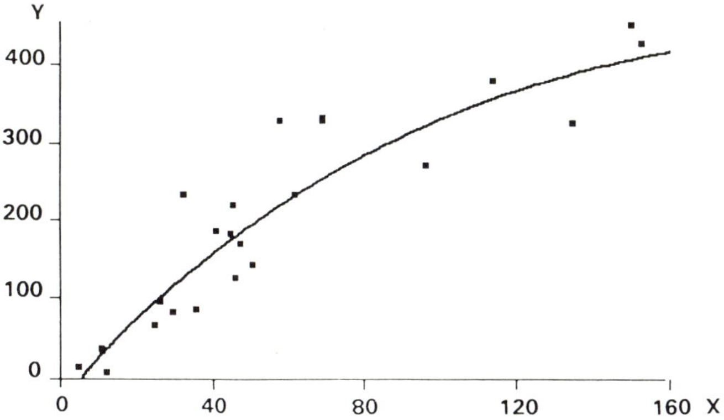

Fig. 8.11. Number of born K_1 cohorts as a function of population size in April. Regression formula: $Y = -23.33 + 4.99X - 0.0139\,X^2$, $R = 0.937$, $n = 22$, $F = 68.86$, $p < 0.0001$. *Y* Number born; *X* population numbers in April

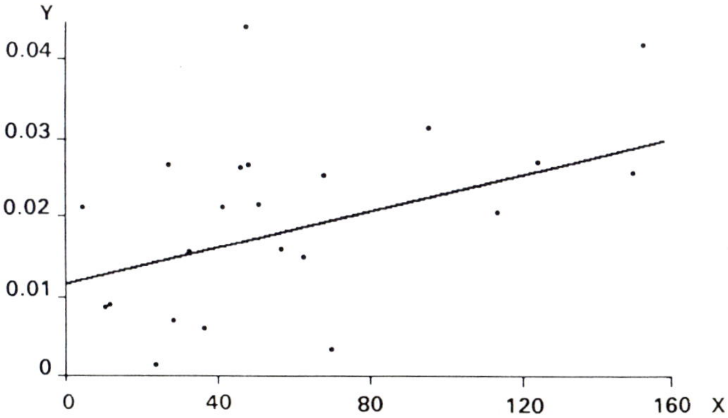

Fig. 8.12. Instantaneous mortality rate of nestlings K_1 as a function of population size in April. Regression formula: $Y = 0.01259 + 0.0001121X$, $r = +0.441$, $n = 22$, $F = 4.84$, $p < 0.02$. *Y* Mortality rate; *X* population numbers in April

(i.e. during the nestling period and first days after weaning). Of these four factors only the number of overwintered that died appears insignificant, and the number of those born most affects the rate of population growth (Table 8.1). Nestling mortality also strongly affects the growth of the population (Table 8.1). The relationship between the number born, K_1 and the initial population numbers (Fig. 8.11) resembles that between the population size in June and in April (Fig. 8.9). The mortality rate of K_1 in the nestling period is linearly correlated with the initial population size (Fig. 8.12), although the data scatter is remarkably high.

8.4 Discussion

The problem regarding the minimum viable population has practical implications. Changes in the habitat structure and quality due to human activities often result in a decrease in population size and/or population area. Which properties of the population and of its habitat are important for its viability is the question that presents a real challenge to contemporary ecology.

The minimum population size depends on biological properties of the species involved: its net reproduction rate, social organization and the power of dispersal. Moreover, the species plasticity or adaptability to new and extreme conditions, although commonly understood, is rarely investigated and usually seen at the moment that such conditions are met.

The bank vole is a wide spread species in Europe. In northern Fennoscandia, cases of population depression were described in this species, particularly in spring (Hansson and Henttonen 1985). This is followed by a population growth lasting 2–3 years. The coefficient of fluctuation (Whittaker 1975) is high, and does not decrease as the breeding season proceeds (Bujalska, in press a). On the other hand, in the central part (optimum) of the species' geographical range, the values of the coefficient of fluctuation decrease from spring to autumn (Bujalska 1985b, 1988). One can suppose that in various sites occupied by the bank vole either different population mechanisms are acting or their action has variable intensity. It seems to be possible that the duration of the breeding season is an important factor that provides different conditions for population growth in the northern and central parts of the species range (Bujalska, in press b).

The results presented in this chapter, primarily the relation of the rate of population growth and the number of young produced to the initial population size, can be interpreted in two ways, either in terms of acceleration of population growth at a low size of the initial population, or in terms of inhibition of growth that accompanies a high initial population size. In both cases, territoriality of sexually mature females is the basic reason for the observed changes in demographic parameters. Female territoriality results in the limitation of the number of sexually active females in the high initial population. This simple mechanism acts automatically, and is often interpreted as a mechanism of population regulation.

Therefore, the key factor for population growth in bank voles is the initial population size: it determines the number born, K_1. Mortality of the cohort K_1 is highly variable, as it depends not only on the intrinsic factor, i.e. mortality of the mothers (Bujalska 1985a), but supposedly also on the action of unpredictable extrinsic mortality factors, like predators that may be present in high or low quantities. The ultimate factor that prevents extinction of this island population is the immigration of strange individuals. However, immigration of individuals from the mainland does not seem to be a frequent phenomenon.

Acknowledgements. Hermann Remmert suggested the topic of this paper and critically reviewed the first draft. Horst Korn also made valuable comments on the manuscript. Stimulating remarks were also provided by participants of the symposium on Kleinstmögliche Population von Pflanzen und Tieren, held in Bad Homburg in 1991.

References

Boer PJ den (1968) Spreading of risk and stabilization of animal numbers. Acta Biotheor 18: 165–194

Boer PJ den (1990) The survival value of dispersal in terrestrial arthropods. Biol Conserv 54: 175–192

Bujalska G (1970) Reproduction stabilizing elements in an island population of *Clethrionomys glareolus* (Schreber, 1780). Acta Theriol 15: 381–413

Bujalska G (1975) Reproduction and mortality of the bank voles and the changes in the size of an island population. Acta Theriol 20: 41–56

Bujalska G (1983) Reproduction. In: Petrusewicz K (ed) Ecology of the bank vole. Acta Theriol 28(1): 148–161 (Suppl)

Bujalska G (1985a) Fluctuations in an island bank vole population in the light of the study on its organization. Acta Theriol 30: 3–49

Bujalska G (1985b) Population dynamics of an island population of *Clethrionomys glareolus*. Acta Zool Fenn 173: 29–34

Bujalska G (1988) Life history consequences of territoriality in the bank vole. In: Boyce MS (ed) Evolution of life histories of mammals. Yale University Press, New Haven, pp 75–90

Bujalska G (in press a) Population dynamics in *Clethrionomys glareolus* (Schreber, 1780). Proc 1st Eur Congr of Mammalogy

Bujalska G (in press b) Population growth of *Clethrionomys glareolus* (Schreber, 1780): consequences of a model of social organization. Pol Ecol Stud 17

Bujalska G, Grüm L (1989) Social organization of the bank vole (*Clethrionomys glareolus*, Schreber 1780) and its demographic consequences: a model. Oecologia (Berl) 80: 70–81

Bujalska G, Mieszkowska D (1984) Distribution of individuals and captures in an island population of the bank vole. Acta Theriol 29: 147–158

Bujalska G, Ryszkowski L (1966) Estimation of the reproduction of the bank vole under field conditions. Acta Theriol 11: 351–361

Bujalska G, Andrzejewski R, Petrusewicz K (1968) Productivity investigation of an island population of *Clethrionomys glareolus* (Schreber, 1780). II. Natality. Acta Theriol 13: 415–425

Hansson L, Henttonen H (1985) Regional differences in cyclicity and reproduction in *Clethrionomys* species: are they related? Ann Zool Fenn 22: 277–288

Jolly GM(1965) Explicit estimates from capture-recapture data with both death and dilution stochastic model. Biometrika 52: 225–247

Pucek M (1983) Habitat preference. In: Petrusewicz K (ed) Ecology of the bank vole. Acta Theriol 28(1): 31–40 (Suppl)

Raczyński J (1983) Taxonomic position, geographical range and the ecological distribution. In: Petrusewicz K (ed) Ecology of the bank vole. Acta Theriol 28(1): 3–20 (Suppl)

Traczyk H (1965) Roślinność "Wyspy Dzikiej Jaboni" na Jeziorze Bełdańskim. Fragm Florist Geobot 11: 541–545

Traczyk H (1971) Relation between productivity and structure of the herb layer in associations on 'The Wild Apple-Tree Island" (Masurian Lake District). Ekol Pol 19: 333–363

Whittaker RH (1975) Communities and ecosystems, 2nd edn. MacMillan-Collier, New York

Zejda J (1966) Litter size in *Clethrionomys glareolus* (Schreber 1780). Zool Listy 15: 193–206

9 Minimum Populations of Carabid Beetles (Col., Carabidae)

L. Grüm

9.1 Introduction

Intuitively, the minimum viable population represents the size that ensures persistence, i.e. in such a population the loss due to mortality, and possibly emigration, can be counterbalanced by reproduction. This concept, stressing the importance of size, seems to be valid, however, for isolated populations as well as for metapopulations, because one can neglect immigration of strange individuals to the population.

Carabid species seem to form populations divided into distinguishable local units (subpopulations, interaction groups) occupying favourable habitats, and dispersal of the beetles is considered an important factor for persistence of local population units (den Boer 1968, 1990).

In the breeding season, solitary beetles roam within the local unit area (Grüm 1971), and their walking behaviour is strongly affected by food (Grüm 1966, 1971; Baars 1979; Mols 1979) and habitat conditions (Grüm 1971). Females may be mutually repulsive, which results in a distribution pattern less clumped than that of males (Grüm 1990). Therefore the breeding success may depend on the rate of male to female encounters, especially on copulation being frequently repeated. The frequency of male to female encounters will depend on the mean beetle density rather than on the total number of the beetles in the local population unit. Among other factors influencing frequency of the encounters one can list sex ratio and running activity.

The aim of this chapter is to estimate synchrony in the seasonal distribution of male and female activity, as well as co-occurrence of males and females in space, and to find out their relations to the mean population density in the breeding season. Additionally, an attempt to evaluate the importance of mean local population density for population persistence will be undertaken.

9.2 Area and Methods

Data were sampled in various forest habitats (Circaeo-Alnetum, Tilio-Carpinetum, Pino-Quercetum and Vaccinio myrtilli-Pinetum) of the Kampinos National Park (located some 17 km NW of Warsaw, Poland). Each sample consists of beetles caught in a set of pitfalls, inspected every few days during a period of

6–8 weeks centred around the middle of the breeding season of the species in question. The sampling took place in the years 1958–1966, 1972–1973 and 1984–1986.

The material consists of individuals of the carabid species with a dry body weight over 15 mg. They belong to the following genera: *Carabus* (nine sp.), *Cychrus* (one sp.), *Calosoma* (one sp.), *Nebria* (one sp.), and *Pterostichus* (five sp.). The remaining carabid species of lesser body weight were excluded because pitfall trapping appears to be inappropriate for assessment of their numbers (Greenslade 1964). Particular attention was paid to elaborate data for the species *Carabus* for which the following estimates were made:

1) General timing of running activity of beetles of each sex. The so-called mean emergence day was calculated for each species, sex and sample. The mean emergence day (MED) was a day of the year (the days being consecutively numbered from 1 to 365) on which 50% of the beetles of the given species, sex and sample was caught. The MEDs for males and females were then correlated using the parametric linear regression method.

2) Synchrony of short-term outbursts of running activity of males and females. The synchrony was assessed for separate samples (each sample consisted of several trap inspections) and species using two methods: Spearmann rank correlation (for more numerous samples) or 2 × 2 contingency tables and the Chi-square test for independent samples.

3) Spatial co-occurrence of males and females. This was assessed for separate samples (each sample was obtained with a set of pitfalls) and species, and each pitfall inspection was a sampling unit. As a result, the numbers caught by such a sampling unit were usually low; consequently, 2 × 2 contingency tables and the Chi-square test for independent samples were applied to infer the co-occurrence

4) The index of mean population density. Because all the *Carabus* species exhibit similar running activity, i.e. the mean shift of the individual position in space varies within the limits of 6–13 m per 24 h (Grüm 1984), it is possible to consider the mean number of beetles caught per pitfall per day as indicative of the mean density. To calculate the mean population density, a formula by Kudrin (1965) was used; however, the results were expressed in terms of the number of m^2 per beetle (i.e. the inverse of the density).

9.3 Results

The total number of samples obtained for the studied *Carabus* is shown in Table 9.1. The MEDs of males and females, obtained for various samples and species of *Carabus* (Fig. 9.1), are generally well correlated ($r = +0.977$, $n = 49$, $p < 0.0001$). However, the species that reproduce in summer and early autumn, i.e. the so-called autumn breeders (Larsson 1939), seem to exhibit less departures from ideal concordance in timing of the MEDs than those that breed in spring (Fig. 9.1). It has turned out that the difference (expressed in days) between the

Table 9.1. Total number of samples for various species of *Carabus*

Development type	Species	No. of samples
Spring breeders	*Carabus arcensis* Hbst.	10
	C. clatratus L.	3
	C. granulatus L.	7
	C. nemoralis Müll.	6
Autumn breeders	*C. coriaceus* L.	6
	C. glabratus Payk.	6
	C. hortensis L.	10
	C. violaceus L.	1

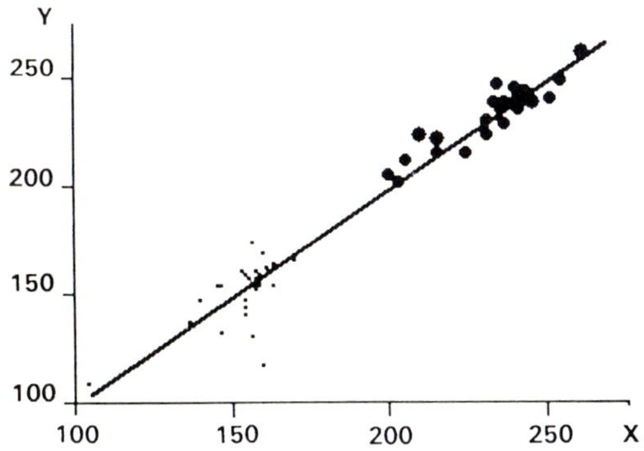

Fig. 9.1. Linear regression of the mean emergence day of males (*Y*) on that of females (*X*) in various samples and species of carabid beetles. *Small* and *large dots* represent spring breeders and autumn breeders, respectively. *Continuous line* is the regression line: $Y = -4.925 + 1.023X$, $n = 49$; $r^2 = 0.955$; $p < 0.0001$. *Numbers on the Y and X axis* represent consecutive days of the year

MEDs calculated for males and females are not correlated with the mean population density index.

The short-term changes in the running activity of males and females are positively correlated in the samples of relatively high index of the mean population density (Table 9.2). However, those changes in cases of relatively low density index are in most cases uncorrelated (Table 9.2). The mean number of beetles per pitfall per day is significantly higher in the samples showing synchrony in running activity of males and females than that in the remaining samples: the one-tailed Mann-Whitney U-test provided $p = 0.0006$ for the data set presented in Table 9.2.

It has proved that positive (non-random) spatial co-occurrence of males and females is characteristic of populations exhibiting relatively high density indices (Table 9.3). In the case of low population density, male and female co-occurrence in space seems to be a purely random event (Table 9.3). The mean number of beetles per pitfall per day is significantly higher in the samples showing positive spatial co-occurrence of males and females than in the samples with random encounters between the beetles of the opposite sex: the one-tailed

Mann-Whitney U-test, applied to the data presented in Table 9.3, provides $p = 0.0121$. In the remaining samples of the number of beetles per pitfall per day below the lowest value shown in Tables 9.2 and 9.3 (i.e. below 0.006), males and females showed only random encounters (Chi-square $p > 0.05$ in all instances).

On the assumption that the mean distance passed by the average beetle per day equals 10 m, and making use of the formula by Kudrin (1965), one can roughly calculate that the highest value of the numbers caught per pitfall per day shown in Table 9.3, i.e. 0.134, is the equivalent of one beetle per $5.2\,m^2$, and the lowest value, i.e. 0.006, represents one beetle per $116.6\,m^2$.

Table 9.2. Spearman rank correlations of changes in running activity of males and females during the breeding season

Species	Sample Symbol	Spearman correlation r	p	Mean no. of beetles per pitfall per day
C. arcensis	K63	+0.65	<0.01	0.134
C. arcensis	S63	+0.49	<0.05	0.101
C. arcensis	G63	+0.39	<0.05	0.034
C. arcensis	A56	+0.44	<0.01	0.021
C. arcensis	A57	+0.58	<0.01	0.018
C. granulatus	A56	+0.57	<0.01	0.015
C. hortensis	A56	+0.14	>0.05	0.009
C. arcensis	A58	+0.33	>0.05	0.008
C. hortensis	A57	−0.05	>0.05	0.008
C. hortensis	A60	+0.13	>0.05	0.007
C. arcensis	A59	+0.11	>0.05	0.006
C. arcensis	A60	−0.06	>0.05	0.006
C. hortensis	A58	+0.24	>0.05	0.006

Table 9.3. Spatial co-occurrence of males and females in pitfalls during the breeding season

Species	Sample symbol	Chi square value	p	Mean no. of beetles per pitfall per day
C. arcensis	K63	33.7	<0.001	0.134
C. arcensis	S63	15.3	<0.001	0.101
C. arcensis	G63	10.4	<0.002	0.034
C. arcensis	A56	0.48	>0.5	0.021
C. hortensis	A56	0.45	>0.5	0.009
C. arcensis	A58	0.01	>0.9	0.008
C. hortensis	A57	6.10	<0.02	0.008
C. hortensis	A60	0.23	>0.6	0.007
C. arcensis	A59	0.11	>0.7	0.006
C. arcensis	A60	0.20	>0.6	0.006
C. hortensis	A58	0.22	>0.6	0.006

Table 9.4. Changes in carabid species composition in a forest habitat

Species	1958–66	1972–73	1984–86
Carabus arcensis Hbst.	+	+	+
Pterostichus niger (Schall.)	+	+	+
P. oblongopunctatus (F.)	+	+	+
P. vulgaris (L.)	+	+	+
C. hortensis L.	+	+	+
C. glabratus Payk.	+	+	+
C. nemoralis Müll.	+	+	+
C. granulatus L.	+	+	+
P. anthracinus (Illig.)	+	+	+
C. clatratus L.	+	+	+
Nebria brevicollis (F.)	–	–	+
Cychrus caraboides (L.)	+	+	+
C. violaceus L.	+	–	–
C. coriaceus L.	+	+	+
C. cancellatus Illig.	+	+	+
Calosoma inquisitor (L.)	–	+	+
P. aterrimus (Hbst.)	+	–	–

Species shown in descending order of the numbers caught, + denotes presence and – absence in pitfalls

In a forest habitat for which the data on the presence of carabid species were available for the years 1958 to 1986, it can be shown that out of 15 species present at the end of the 1950s and the beginning of the 1960s two species have never been caught since the beginning of the 1970s (Table 9.4). However, two other species have appeared (Table 9.4). The mean population densities of the extinct species, as well as the densities of the invaders, are among the lowest (Table 9.4).

9.4 Discussion

Chance events in genetic, demographic and environmental variables may affect the probability of extinction of a small population (Schaffer 1987; Soule 1987). This is the main axiom of the MVP theory, which points out the properties resulting from the population size. A different approach is presented in the theory of spreading of risk (den Boer 1968, 1990): population viability results from heterogeneity of both the population and its environment rather than from population size. This difference seems to stem from different population theories: single population (former) and composite population (latter).

The empirical data presented in this chapter indicate that in the local populations of the carabid beetles studied (genus *Carabus*) there is, in general, a significant correlation between timing of breeding periods of males and females. However, the short-term changes in the running activity of males and females during the breeding season are synchronous only in relatively dense populations.

Also, the co-occurrence of males and females in the same pitfalls (representing small sections of the inhabited area) is higher than expected from chance events only in the case of populations of relatively high density.

One could speculate that relatively low densities in carabid populations disfavour breeding success because of the low chance of repeated copulation. A set of data on long-term monitoring of species composition of a carabid community is in accordance with the hypothesis that in populations of low density the breeding success is unsatisfactory: the species that have locally become extinct belonged to the least numerous. On the other hand, however, the species that have recently successfully invaded the habitat under study also exhibit low mean values of local population density. Hence, low population density (or size) may not be considered as the decisive factor for population persistence in the case of carabid species. Further studies aimed at solving the problem of viability of small (or of low density) populations will need more information on both demographic parameters in small populations (extinct and invading), as well as on the dispersal of individuals and the rate of their exchange between local units (interaction groups or subpopulations).

References

Baars M (1979) Patterns of movement of radioactive carabid beetles. Oecologia (Berl) 44: 125–140

Boer PJ den (1968) Spreading of risk and stabilization of animal numbers. Acta Biotheor 18: 165–194

Boer PJ den (1990) The survival value of dispersal in terrestrial arthropods. Biol Conserv 54: 175–192

Greenslade PJM (1964) Pitfall trapping as a method for studying populations of Carabidae (Coleoptera). J Anim Ecol 33: 301–310

Grüm L (1966) Diurnal activity rhythm of starved Carabidae. Bull Acad Pol Sci Cl II 14: 405–411

Grüm L (1971) Spatial differentiation of the *Carabus* L. (Col., Carabidae) mobility. Ekol Pol 19: 1–34

Grüm L (1984) Carabid fecundity as affected by extrinsic and intrinsic factors. Oecologia (Berl) 65: 11–121

Grüm L (1990) Spatial distribution of males and females of *Carabus arcensis* Hbst. in the breeding season. In: Stork NE (ed) The role of ground beetles in ecological and environmental studies. Intercept, Andover, pp 277–287

Kudrin AI (1965) K voprosu o primenenii zemlianykh lovushek dlia izuchenia raspredelenia i vzaimodeistvia elementov entomofauny na poverkhnosti pochvy. Tr Vses Entomol Ova 50: 272–290

Larsson SG (1939) Entwicklungstypen und Entwicklungszeiten der Dänischen Carabiden. Entomol Medd 20: 277–560

Mols PJM (1979) Motivation and walking behaviour of the carabid beetle *Pterostichus coerulescens* L. at different densities and distributions of the prey. Misc Pap Landbouwhogesch Wageningen 18: 185–198

Schaffer ML (1987) Minimum viable populations: coping with uncertainty. In: Soulé ME (ed) Viable populations for conservation. Cambridge University Press, Cambridge, pp 69–86

Soulé ME (ed) (1987) Viable populations for conservation. Cambridge University Press, Cambridge

10 To Survive or To Become Extinct: Small Populations of Tetraonids in Central Europe

S. KLAUS

10.1 Introduction

Population size has been considered to be the primary determinant of the likelihood of extinction in many studies, providing empirical support for the minimum viable population size (MVP) theory (Shaffer 1981; Ryan and Siegfried, Chap. 2, this Vol). In many cases, however, loss of genetic heterogeneity, inbreeding depression, chance demographic events and—possible the most important factor—environmental variability play a major role in the extinction of small populations. In this chapter, however, examples for small tetraonid populations with differing persistence are summarized. In a few cases, very small populations of black grouse or capercaillie started to increase after many years of depression.

Hazel Grouse (*Bonasa bonasia*), capercaillie (*Tetrao urogallus*) and to a lesser extent black grouse (*Tetrao tetrix*) are extremely resident birds. Because of behavioural (small home ranges, year-long residents), morphological (low wing area/body mass ratio, and low heart/body mass ratio) and physiological constraints (low energy food), they rarely move over large distances. Therefore, the fragmentation of their habitats by human activities has caused splitting of the formerly closed Eurasian forests into habitat islands and isolated subpopulations. This process is especially pronounced in Central Europe. On the other hand, also in natural landscapes of the northern boreal taiga, these species are often distributed patchily because of their dependence on different successional stages of forests: the black grouse as a pioneer species inhabits the youngest stages of forest succession, immediately following burning, windfall, avalanches and bark beetle calamities. The hazel grouse prefers young forest of an age of 10–50 years or unevenly aged old forests, whereas the capercaillie inhabits the oldest stages of forests with openings and rejuvenation islands (Swenson and Angelstam 1993).

According to our own studies and a review of the literature, these three species seem to be able to survive quite long periods of time even in small isolated or semi-isolated populations. This is true for natural habitats as well as forests fragmented by human influences. Sometimes, increases have been observed after long-term stagnation of small, isolated populations, when the ecological situation has become more suitable. On the other hand, there are

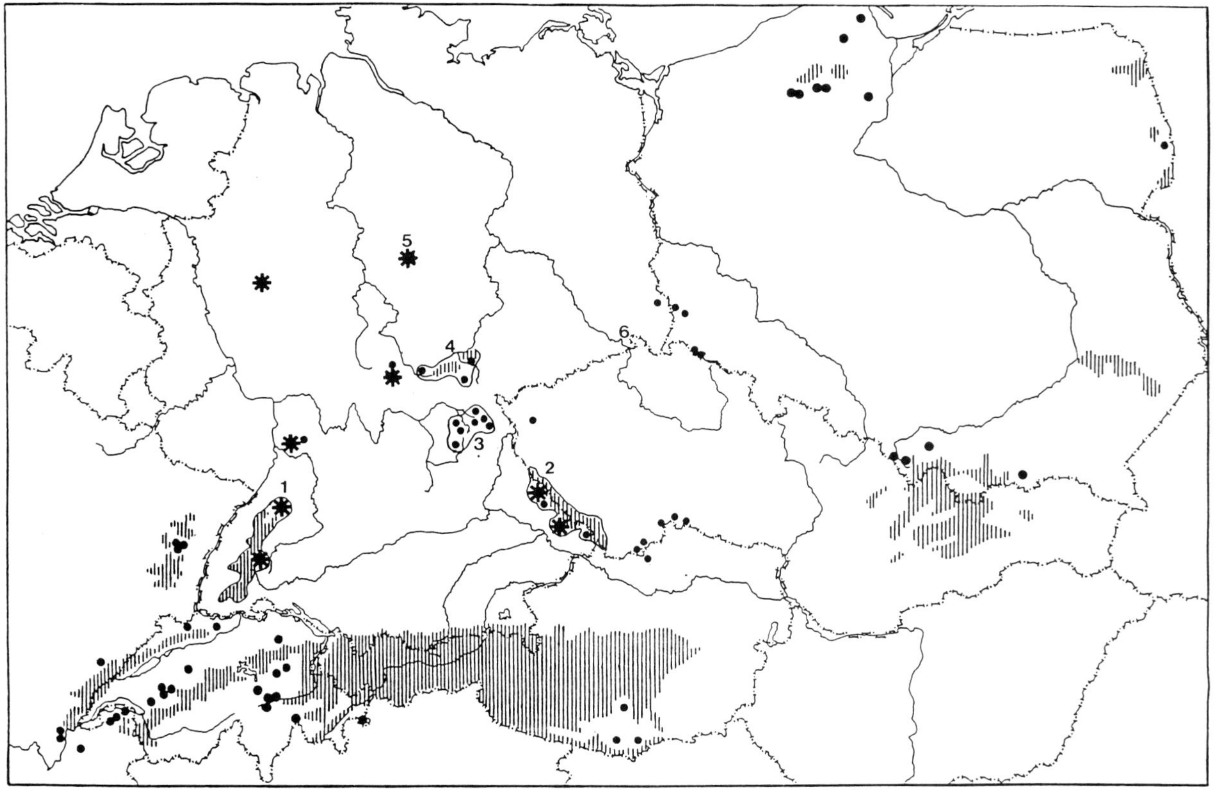

Fig. 10.1. Capercaillie distribution in Central Europe in 1992 (Klaus et al. 1989 and unpubl. results). Isolated populations mentioned in this chapter are indicated by numbers: *1* Black Forest; *2* Bavarian/Bohemian Forest; *3* Fichtelgebirge; *4* Thuringia; *5* Harz; *6* Saxonian Switzerland (extinct since 1975)

also examples of rapid extinction of small, isolated populations of tetraonids. In most cases, the causes of the extinction process have not been analyzed.

Some examples of declining or stable small populations of grouse will be discussed here. Knowledge about the minimum viable population size in tetraonids is lacking and only indirect conclusions can be drawn from our case studies.

10.2 Capercaillie

The capercaillie, the largest and most dimorphic grouse, inhabits boreal coniferous forests across the northwestern and central palaearctic region.

This species has relatively small home ranges (adult cocks: 50 ha, Müller 1974; 10–78 ha, Wegge and Larsen 1987; hens: 40 ha, Müller 1974, 14–38 ha, Gjerde (et al. 1984). By radio tagging, Wegge and Larsen (1987) found that young cocks visited several leks in an area, moving up to 15 km. In their third year, they became resident, and movements from and to the lek were not longer than 4 km (for a review, see Klaus et al. 1989). Hens moved 0.6–6.6 km (mean 2.1 km) between the lek and the nesting site. According to Rolstad (1989), mean distances between home lek and summer occupancy centres did not differ between age groups of males. For individuals of 2 years and older, these distances varied between 80 m and 10 km (mean 1.47 km). Also, in the Russian taiga (NW Ural), more than 90% of ringed birds were retrapped less than 4 km away from their ringing site (Romanov 1988). Only a few percent of the population migrated over 20–40 km (Glutz von Blotzheim et al. 1973).

From these findings it can be concluded that most of the small surviving populations in Central Europe (Fig. 10.1) are isolated genetically. In Germany, north of the Alps, there are several isolated populations of different size (Black Forest: 1000 birds, Bavarian Forest: 60, Fichtelgebirge: 60, Thuringian Forest: 30). A reintroduced population, created by releasing birds reared in captivity, inhabits the higher parts of the Harz Mountains (Haarstick 1979, 1992). As shown in Table 10.1, smaller populations decrease more rapidly than larger ones. An interesting exception to this rule is the population in the Fichtelgebirge

Table 10.1. Half-life times (tH) of declining capercaillie populations

Area	Population size	(year)	tH	Reference
Swiss Alps[a]	1100 (1970), 600	(1986)	16	Marti (1986)
N Black Forest[a]	250 (1971), 125	(1983)	12	Roth and Suchant (1990)
Bohemian Forest	300 (1975), 150	(1985)	10	Kučera (pers. comm.)
Thuringia	300 (1970), 150	(1978)	8	Klaus et al. (1985)
Thuringia	80 (1985), 40	(1989)	4	Klaus (1991b)
Saxonia[b]	32 (1960), 16	(1963)	3	Klaus and Augst (unpubl.)

[a] Displaying cocks.
[b] Saxonian Switzerland only.

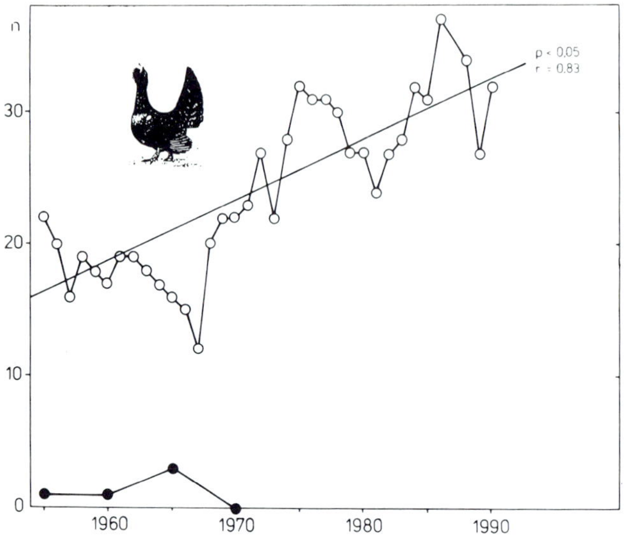

Fig. 10.2. Number of displaying cocks (○) and cocks shot (●) in the Fichtelgebirge. (Bachmann 1991)

Fig. 10.3. Capercaillie habitats in the Fichtelgebirge are characterized by immission-caused openings in the formerly closed spruce forests with mosaic-like distribution of grass and bilberry. (Photo S. Klaus)

(Bachmann 1991; Hertel, Pers. comm; Braun, pers. comm.), where a small isolated population has increased in the last 20 years, probably due to the creation of an optimal forest structure resulting from forest damage (Figs. 10.2, 10.3). The former closed and dark spruce plantations at high elevations were changed by immission-caused forest damage to a mosaic of patches of old forest, rejuvenation stands and openings with grasses and/or bilberry (*Vaccinium myrtillus*). The present situation is very labile. This example shows that a small population of less than 40 adults who were isolated for a long time was able to multiply after a bottleneck situation when the habitat quality increased.

A contrary development was observed in Thuringia. As shown in Fig. 10.4, the size of this population decreased from about 300 adults in 1970 to only 30 in 1990. In the same time, the area of distribution declined from about 1200 km^2 to only 300 km^2. The area declined in a nearly classical way: capercaillie disappeared at first from the periphery of their range. The main cause of the decline was assumed to be reduction in the amount of old forest by clear-cutting (Figs. 10.5, 10.6). This resulted in a fragmentation of the most valuable parts

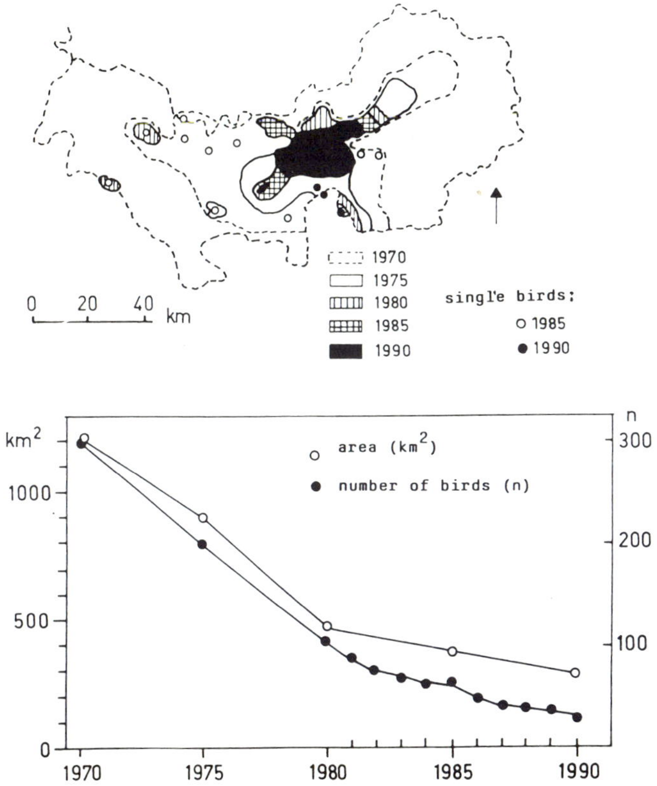

Fig. 10.4. Shrinking of the Thuringian capercaillie area (*top*) in 5-year intervals. *Bottom* Decline of adults in spring (●) and shrinking of the area size (○, km^2)

Fig. 10.5. Old mixed forests with pine and spruce formed optimal habitats for capercaillie in Thuringia. (Photo H. Böttinger)

of the habitats (Klaus 1991a). A comparison of demographic data of periods with high (about 240 birds in 1973) and low population size (75 birds in 1982) is shown in Table 10.2. Mean clutch size and hatching success did not differ between the two periods. In contrast, predation, especially nest predation, was much higher in the period of low population size (Klaus 1984). The change in the sex ratio in favour of hens was probably caused by habitat deterioration and/or increase in human disturbances (Wegge 1980).

10.3 Black Grouse

The black grouse is considered to be more mobile than hazel grouse or capercaillie. Otherwise, home ranges are rather small (20–100 ha). In the Alps (Aletsch reserve), Marti (1985) found very small winter home ranges: a group of ten cocks used only 105 ha for longer time periods. The dispersion of juvenile black grouse was studied by Willebrand (1988) using radio tagging. Hens dispersed 6.3 km (mean) from their place of birth. Cocks were more resident than females. The longest movements of ringed birds were 20–26 km in Scandinavia (Glutz von Blotzheim et al. 1973). Especially in spruce forest

Fig. 10.6. Habitat deterioration by clear-cutting in the Thuringian Slate Mountains. (Photo S. Klaus)

Table 10.2. Comparison of demographic data between two periods with high (1971–1975) and low (1976–1985) abundance of capercaillie

Period	1971–1975		1976–1985	
Sex ratio on the leks (♀/♂)	1.1 ± 0.1	(n = 18)	1.8 ± 0.2	(n = 20)[a]
Mean clutch size	7.6 ± 1.3	(n = 14)	7.5 ± 0.5	(n = 13)
Hatching rate (%)	94.6 ± 7.4	(n = 9)	91.3 ± 12.7	(n = 7)
Chicks hatched per successful nest	7.2 ± 0.9	(n = 9)	7.2 ± 0.4	(n = 7)
Sex ratio in broods (♀/♂)	1.0 ± 0.2	(n = 8)	2.2 ± 0.8	(n = 11)
Nests destroyed (%)[b]	35.3		66.7	
Daily mortality coefficient of chicks (no./day)	0.013 ± 0.007		0.018 ± 0.009	

[a] Calculated for the period 1980–1985.
[b] Cause of nest predation: wild boar (31%), mammalian predator (red fox, marten, badger 44%), corvids (6%), human disturbance (19%).

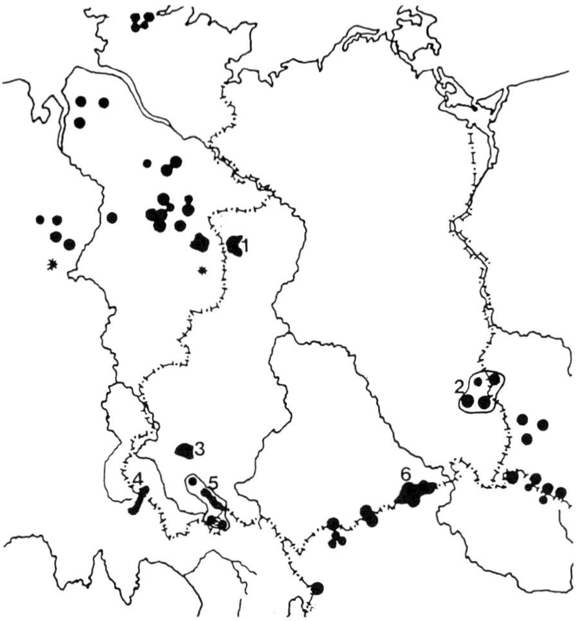

Fig. 10.7. Distribution of black grouse in the former German Democratic Republic and adjacent parts of Lower Saxonia, Bavaria, Bohemia and Poland (Klaus 1991b). Populations separated by more than 25 km space are indicated by numbers: *1* Colbitz-Letztlinger Heide; *2* Muskauer Heide near Weißwasser; *3* military area Ohrdruf; *4* Rhön; *5* Thuringian Forest; *6* eastern part of the Erzgebirge

habitats in the mountains, black grouse is an extremely resident bird (Porkert, pers. comm.). From these findings, it can be concluded that black grouse populations can also become isolated if the distance to other subpopulations is more than 30 km.

Examples of small isolated black grouse populations in eastern parts of Germany are shown in Fig. 10.7. The eastern part of the Erzgebirge (40–60 birds), Rhön (40–60 birds), Thuringian Forest (20 birds), and some military areas (Ohrdruf/Thuringia: 30 birds; Colbitz-Letztlinger Heide: 30–40 birds; Muskauer Heide: about 100 birds; Fig. 10.8).

In the Thuringian Forest, an interesting development was observed after 1946. A hurricane-force wind caused openings in the spruce forests along the crest of this mountain range (20 000 ha in total) followed by bark beetle calamities and fire. Formerly, black grouse occurred in very low densities only near peat bogs and meadows, isolated for decades from other populations. Afforestation started with rowan, *Sorbus aucuparia*, and birch, *Betula pubescens* and *B. carpathica*. After some years, large areas were covered with young forests of these species, islands of regenerating spruce and extended open fields with grass and bilberry (Fig. 10.9). Figure 10.10 shows the increase in black grouse numbers

a

b

Fig. 10.8a, b. Black Grouse habitats on military areas often involve different successional stages of forests grown up after burning. **a** Ohrdruf/Thuringia. **b** Muskauer Heide. (Photo S. Klaus)

Fig. 10.9. Black grouse habitats on the crest of the Thuringian Forest in 1952, 6 years after the catastrophic windfall followed by bark-beetle calamities. (Photo E. Seibt)

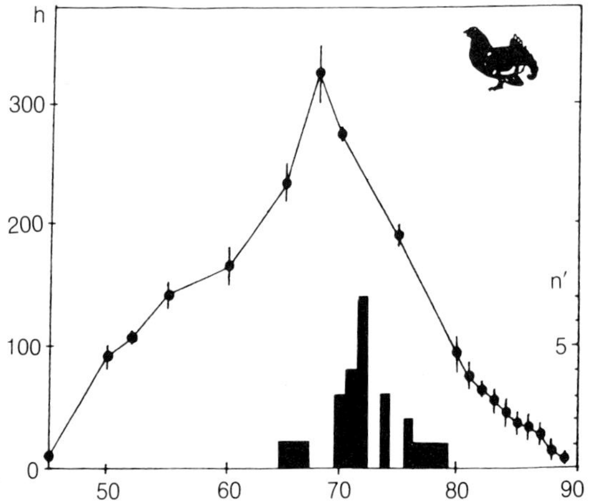

Fig. 10.10. Numbers (range) of cock and hen black grouse (●) counted on leks in the Thuringian forest (1945–1989, n, left axis) and the number of black grouse killed by predators (black columns, n, right axis). (Klaus et al. 1991)

from about 10 in 1945 to more than 300 in 1968. At the same time, dispersal of black grouse was observed. Whereas the birds lived in an area of about 4000 ha in 1945, in 1968 they used an area of about 120 000 ha including fields, meadows and clearcuts throughout the lower Thuringian Forest and the Thuringian Slate Mountains.

Planting of spruce started in 1953. Because of the harsh climatic conditions on the crest of the mountains, spruce grew slowly and therefore optimal habitats

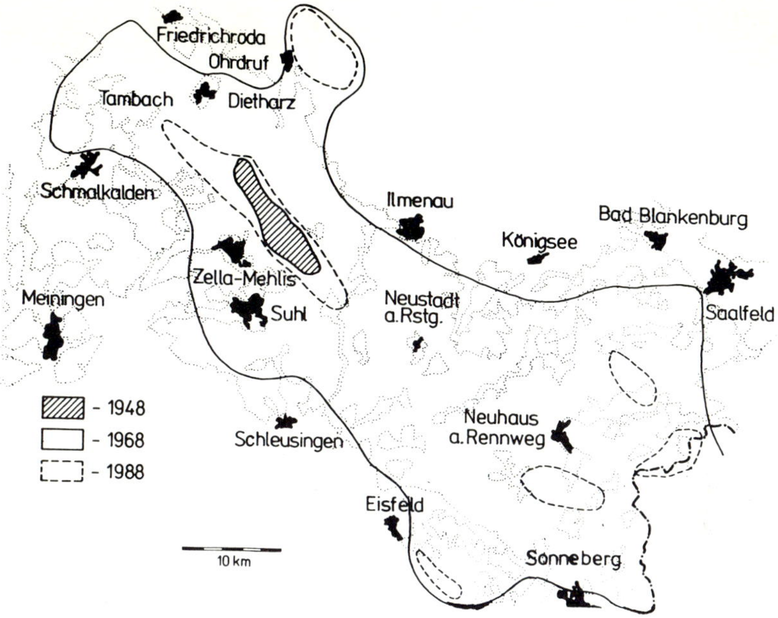

Fig. 10.11. Extension and shrinking of the Thuringian Black Grouse area (Klaus et al. 1991): *hatched* — area size in 1948; — area during the population maximum in 1968; ----- centres of recent distribution

for black grouse were preserved over several decades. Later, birch and rowan, the main winter foods, were cut and canopies in spruce plantations became closed. After 25 years of continuous increase, a rapid decrease in black grouse numbers occurred, resulting in a population minimum in 1990 that was similar to that in 1945.

Once again, the peat bogs are the centres of black grouse distribution. Predation (as measured by located kills) increased greatly after 1965, when the forest canopy closed. Figure 10.11 shows the extension of the range after 1946 and its retraction after 1968.

The situation in 1990 is characterized by a shortage of winter food in the spruce monocultures. In addition, bilberry has been over-grazed by the dense population of red deer, *Cervus elaphus*. Air pollution and increase of tourism have also limited the bird's use of habitat. For these reasons, black grouse are unable to start a new cycle of increase today even though modern forestry practices have created large clearcut areas which are structurally similar to the openings caused by the windfall in 1946 (Klaus et al. 1991).

There are also other examples of increasing black grouse populations starting from small isolated remnants of formerly large populations. In the Erzgebirge, black grouse declined rapidly in the 1970s due to forest damage on the whole crest of the mountains. The replacement of spruce monocultures by young

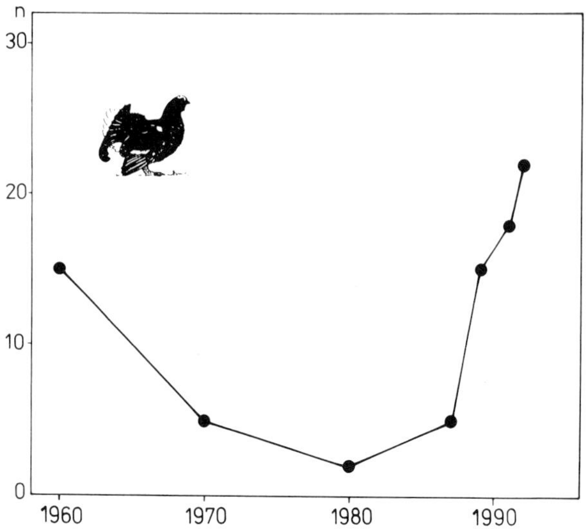

Fig. 10.12. Bottleneck in a black grouse population development: decline and increase in the eastern part of the Erzgebirge. (Schulenburg 1991 and unpubl. results)

Fig. 10.13. Black grouse habitat in the Erzgebirge: after cutting of damaged spruce forests, succession started on extended open areas. (Photo R. Heyer)

growth of birch, rowan, larch and aspen, in combination with openings, grassland and bilberry shrubs, resulted in a constant increase in the population and extension of the area after 1980 (Figs. 10.12, 10.13). Also in some military areas in Austria, Bohemia and Germany, there are small isolated subpopulations of black grouse (between 30 and 200 birds) that have not declined in the course of several decades (Hüppop et al. 1988; Hofmann et al. 1992; Bufka, Cherveny, Hertel, Schmalzer, unpubl.) or which have increased from small founder populations (Brozio 1990).

Additional information on small, increasing populations of black grouse comes from releasing experiments with wild birds. According to Hölzinger (1980) black grouse was extinct in Pfunger Ried in Upper Suebia by 1837/38. In 1889, 18 adult cocks and 36 hens from Sweden were released here. In this same year, 6 broods were observed including 42 chicks. In April 1892, 150 birds were counted. The latest individuals were observed quite recently, in 1967.

In an area called Dornacher Ried, three Swedish males and six hens were released in 1898 and 1899. These were also successful as a source population (Bertsch 1918). Birds were observed as late as 1945. These examples make clear that even small founder populations of released wild birds can result in a

Fig. 10.14. Hazel grouse habitat in the Black Forest: Older stages of forest succession on former fields with rich mixture of deciduous and coniferous trees form centres of the present distribution. (Photo S. Klaus)

population increase and persistence of an isolated population for decades, providing the habitat is suitable.

10.4 Hazel Grouse

In Germany, only two small separated populations of hazel grouse remained—a rapidly declining one in the Black Forest (40–50 pairs, Asch, pers. comm., Fig. 10.14) and a relatively stable one in the Rheinisches Schiefergebirge (about 120 pairs, Schmidt, pers. comm.). A third population, in the Bavarian Forest, is stable (Scherzinger 1976) and continuous with a large population in Bohemia (Klaus 1991a). The hazel grouse is the most resident species among the European tetraonids. About 90% of ringed birds stayed in the closed vicinity (< 1 km) of the marking site (for review, see Bergmann et al. 1982). New results by Swenson (unpubl.), obtained by radio tagging of wild birds at Grimsö (Central Sweden), showed that hazel grouse is unable to move over open fields more than 150 m wide. This makes the hazel grouse an ideal object for the study of island habitats and isolated subpopulations of birds. An artificial population has been released in the Harz Mountains from birds reared in captivity. A limited number of parental birds formed the breeding stock (Bergmann et al. 1991), which means that this population (which reproduces in the wild) could be genetically homogeneous due to the founder effect. This should be monitored carefully in the future with respect to population growth, genetic variability, and parameters of reproduction like egg number, hatching rate, brood size and mortality.

10.5 Conclusions

In capercaillie, the rate of decline appears negatively related to the size of the subpopulation (Table 10.1). On the other hand, the example from the Fichtelgebirge suggests that even a very small subpopulation is able to survive for decades and even multiply if the habitat quality is improving (Fig. 10.2). Data on capercaillie reproduction in our Thuringian study area (Table 10.2) are not in accordance with the assumption that the egg number, fertilization and/or hatching success in a small remnant population declined as a possible result of inbreeding depression.

Also, the black grouse seems to be able to survive over long periods of habitat deterioration in small populations. After such bottleneck situations, the species is able to multiply rapidly, sometimes starting from a very small founder population, if the ecological conditions became favourable. This happens after burning, windfall, insect calamities or even after immission-caused forest damage (Schulenburg 1992). In Thuringia, the black grouse population size increased by a factor of 10 in the course of only 5 years (Klaus et al. 1991). A similar rapid increase was reported by Sperber (1968) after insect calamities in pine forests of the Nürnberger Reichswald. Such an increase is obviously caused by optimal habitat structure and high quality food resulting from burning and/or insect calamities (Klaus et al. 1991).

It appears possible that in rare cases single individuals move longer distances than described in the literature. By this means, genetic exchange between some individuals of isolated populations could have occurred. However, from ringing and radio tagging, it is known that only a few percent of the populations of capercaillie and black grouse move over distances longer than 20 km. Therefore, genetic isolation can be assumed in most of the cases described here. In the most resident hazel grouse, genetic isolation is easily achieved if open fields separate forest islands (Swenson, unpubl.). Even in continuous forests, hazel grouse disperse short distances (Swenson 1991). The development of several subspecies of hazel grouse in Central Europe (Bauer 1960) could also be due to this fact.

All three forest grouse species form isolated populations in Central Europe. There are examples of slow or rapid extinction of small, isolated grouse populations, but also examples of rapid increase of long-isolated small populations after habitat improvement. This gives us cause for concern, but also hope. We can lose our remnant populations very easily. However, if we begin now to improve habitat conditions, it appears possible to regain viable populations from these population remnants.

Acknowledgements. I am grateful to Wiltraud Engländer, Hans-Heiner Bergmann, Wolfgang Scherzinger and Jon Swenson for stimulating discussions, suggestions to the manuscript and help in correcting the English.

This work was supported by the Fonds der Chemischen Industrie.

References

Bachmann J (1991) Das Auerhuhn im Fichtelgebirge am Beispiel des Ochsenkopfes. Diplomarbeit Weihenstephan

Bauer K (1960) Variabilität und Rassengliederung des Haselhuhns (*Tetrastes bonasia*) in Mitteleuropa. Bonn Zool Beitr 11: 1–18

Bergmann H-H, Klaus S, Müller F, Wiesner J (1982) Das Haselhuhn. Ziemsen-Verlag, Wittenberg-Lutherstadt

Bergmann H-H, Elstrodt W, Körner S, Künne H-J (1991) Ein Versuch der Wiederansiedlung des Haselhuhns (*Bonasa bonasia*), Bedingungen und Methoden. In: Bechet G, Erasmy F, Heidt C (Hrsg) Die Lohhecken des Öslings und das Haselhuhn. Seminarbericht Luxemburg

Bertsch K (1918) Das Birkhuhn in Oberschwaben. Ornithol Monatsber 26: 142–144

Brozio F (1990) Die Flachlandpopulation des Birkhuhns (*Tetro tetrix* L.) im Kreis Weißwasser. Abh Ber NaturKundemus Görlitz 64: 93–98

Gjerde I, Wegge P, Petersen O, Solberg GE (1984) Home range and habitat use of a local capercaillie population during winter in SE Norway. In: Lovel T (ed) Grouse. Lovel T, Hudson P, Suffolk, pp 247–260

Glutz von Blotzheim U, Bauer KM, Bezzel E (1973) Handbuch der Vögel Mitteleuropas, Bd 5. Galliformes und Gruiformes. Akademische Verlagsgesellschaft, Frankfurt/Main

Haarstick K-H (1979) Erfahrungen bei der Haltung und Aufzucht von Auerwild (*Tetrao urogallus* L.) in der Aufzuchtstation im Staatlichen Forstamt Lonau/Harz. Z Jagdwiss 25: 167–177

Haarstick K-H (1992) Die Wiedereinbürgerung des Auerhuhns im Harz—ein Beitrag der Niedersächsischen Landesforstverwaltung zum Artenschutz. Naturschutzreport 4: 95–102
Hofmann P, Küssner J, Bellstedt R (1992) Voruntersuchungen zur Bedeutung des Truppenübungsplatzes Ohrdruf in Thüringen als Lebensraum bestandsbedrohter Tierarten. Artenschutzreport 2: 60–66
Hölzinger J (1980) Der Untergang des Birkhuhns *Lyrurus tetrix* in Baden-Württemberg und dessen Ursachen. Beih Veröff Naturschutz Landschaftspflege Bad Württ 16: 123–134
Hüppop O, Walther K, Holzapfel C, Vidal S, Wilkens H (1988) Isolierte Vorkommen des Birkhuhns (*Tetro tetrix*) im Naturschutzgebiet Lüneburger Heide: Verhalten, Gefährdung, Schutz und Pflegevorschläge. Verh Naturw Ver Hamb 30: 319–343
Klaus S (1984) Predation among capercaillie in a reserve in Thuringia. In: Lovel T (ed) Grouse. Lovel T, Hudson P, Suffolk, pp 334–346
Klaus S (1991a) Effects of forestry on grouse populations: case studies from the Thuringian and Bohemian forests, central Europe. Ornis Scand 22: 218–224
Klaus S (1991b) Bestandssituation, Rückgangsursachen und Schutz der Rauhfußhühner in der ehemaligen DDR. Ber Dtsch Sekt Int Rat Vogelschutz 30: 107–116
Klaus S, Boock W, Görner M, Seibt E, Wennrich W (1985) Zur Bestandssituation des Auerhuhns (*Tetrao urogallus*) in Thüringen. Veröff Mus Gera 11: 94–100
Klaus S, Andreev AV, Bergmann H-H, Müller F, Porkert J, Wiesner J (1989) Die Auerhühner. Ziemsen-Verlag, Wittenberg-Lutherstadt
Klaus S, Seibt E, Boock W (1991) Zur Ökologie des Birkhuhns (*Tetrao tetrix*) im mittleren Thüringer Wald. Acta Ornithoecol 2: 211–229
Marti C (1985) Unterschiede in der Winterökologie von Hahn und Henne des Birkhuhns *Tetro tetrix* im Aletschgebiet (Zentralalpen). Ornithol Beob Bern 82: 1–30
Marti C (1986) Verbreitung und Bestand des Auerhuhns *Tetrao urogallus* in der Schweiz. Ornithol Beob Bern 83: 67–70
Müller F (1974) Territorialverhalten und Siedlungsstruktur einer mitteleuropäischen Population des Auerhuhns, *Tetrao urogallus* major C.L. Brehm. Dissertation, Marburg
Rolstad J (1989) Habitat and range use of capercaillie *Tetrao urogallus* L. in south central Scandinavian boreal forests. Dissertation, Ås-NLH
Romanov AN (1988) Gluchar. Moscow (in Russian)
Roth K, Suchant H (1990) Bestandsentwicklung des Auerwilds in Baden-Württemberg. In: Auerwild in Baden-Württemberg—Rettung oder Untergang. Selbstverlag der Landesforstverwaltung Baden-Würtemberg, Stuttgart, pp 17–22
Scherzinger W (1976) Rauhfußhühner. Schriftenr Nationalpark Bayer Wald 2, pp 6–71
Schulenburg J (1992) Zur Bestandsentwicklung des Birkhuhns (*Tetrao tetrix*) in einem immissionsbelasteten Gebiet des östlichen Erzgebirges. Artenschutzreport 1: 47–51
Shaffer ML (1981) Minimum population sizes for species conservation. BioScience 31: 131–134
Sperber G (1968) Die Reichswälder bei Nürnberg—aus der Geschichte des ältesten Kunstforstes. Mitt Staatsforstverw Bayerns, Nürnberg, pp 114–122
Swenson J (1991) Is the hazel grouse a poor disperser? Trans Int Union Game Biol Congr 20: 347–352
Swenson J, Angelstam P (1993) Habitat separation by sympatric forest grouse in Fennoscandia in relation to forest succession. Can J Zool 71: 1303–1310
Wegge P (1980) Distorted sex ratio among small broods in a declining capercaillie population. Ornis Scand 11: 106–109
Wegge P, Larsen BB (1987) Spacing of adult and subadult common capercaillie during the breeding season. Auk 104: 481–490
Willebrand T (1988) Demography and ecology of a black grouse (*Tetrao tetrix* L.) population. Dissertation, Uppsala

Subject Index

Aci-Göl 51pp
adaptation 56pp
afforestation 144
Allee effect 86, 92
Amazonia 105pp
antbird 108
area requirements 40, 107
autumn breeder 132p

bamboo thicket 108
bank vole 119pp
Berlepschia 108
bilberry 140pp
birch 144pp
birthrate 82p, 86pp
bittersalt lake 51pp
black grouse 137, 142pp, 150
body size 11, 42
bottleneck 2, 5, 9, 30, 37, 43, 141, 150
breeding
– range 28
– season 119pp, 131pp
– success 15, 131, 136

campina vegetation 106p
capercaillie 137pp, 150p
Carabid beetles 131pp
Carabus 132pp
carnivore 40p
carrying capacity 30, 71pp, 87pp
catastrophe 7, 44p
census numbers 36, 45
Cichlidae 55pp
colonization 9, 73pp, 81
– success 77
colour morph 58pp
common census 123
competition 86
– contest 86, 91
– scramble 91p
coniferous forest 139
conservation
– areas 44
– biology 17
– planning 115
– programme 30
co-occurence
– spatial 131pp
corridor
– forest 115
Crabapple Island 119pp
cyprinodonts 51

DDT 31
death rate 82, 86pp
deme 42
demographic noise 69pp, 80, 89p
density 26, 28
– dependence 70p, 80pp
– regulation 71, 80
dentition 58
development stage 86
differentiation 51
disease 35
dispersal 106p
– capability 110
– rate 37
dispersion 24
distribution 25pp
– binomial 85pp, 90
– frequency 72
– normal 85p, 91
– patchy 105pp, 137
– pattern 131
– Poisson 85, 91
– probability 73
– quasistationary 72pp
diversity
– genetic 4, 9, 18, 34, 38pp
domestic animal 33
drift
– genetic 4p, 39

ecology
– theoretical 67pp, 95
ecosystem decay 114
egg model 59pp
El Niño 15

Endemic Bird Areas (EBA) 27
environmental noise 68pp, 78, 89pp
Erzgebirge 147pp
evolution 39, 55
– adaptive 39
expressivity 54
extinction
– dynamics 45p
– probability 73pp, 90
– process 3, 6p, 17, 77, 82
– rate 10, 43
– risk 67p, 73pp, 119
– secondary 12
– vortex 4, 16

feeding
– habit 56pp
– type 40
Fichtelgebirge 139p
fitness 4p, 26
fluctuation
– coefficient 124, 128
– demographic 69
– environmental 67pp
flycatcher 108
forest
– habitat 131, 135
– structure 141
Forest Fragments Project 112p
forestry practice 147
founder 5, 9, 38, 45

gene
– flow 13, 61
– pool 38
generalist 105pp
generation 25p
– overlapping 68pp, 93p
– separated 83pp
– time 45
growth
– exponential 72
– logistic 78pp
growth rate 70pp, 119
– density-independent 71
– intrinsic 70pp, 81

habitat
– conditions 121
– detoriation 142p
– ephemeral 109p
– fragment 10p, 12, 28
– fragmentation 113p, 137, 141p
– generalist 107p
– island 137
– patches 105pp
– quality 141, 150
– requirement 12
– size 95
– specialist 107pp
– structure 151
hazel grouse 137, 149pp
heath hen 16
herbivore 40p
herb layer 121p
heterozygosity 4, 9, 35
home range 25, 28, 139
homogamy 56pp
homozygosity 5
hybrids 54

immigration 12p, 44, 81p
– rate 82, 96
inbred line 15
inbreeding 1, 34p, 150
– coefficient 5, 34
– depression 5, 8p, 18, 35
individual number 68pp, 84
– critical 76pp
insect calamity 150
insularization 33
interference
– human 29p
introgression 10
island 1, 13
– biogeographic theory 10, 39
– habitat 150
– population 1, 9
– taxa 28
isolation 23p, 56
– ecological 58
– effect 28, 113pp
– genetic 151
– mechanism 55
– reproductive 56

K-strategist 2

landbird 25pp
life time 78
– mean 72pp, 88pp
litter size 123p
longevity 25

macropopulation 54p
management 23
– conservational 97
mean emergence day (MED) 132
Mendelian inheritance 54
metapopulation 17, 23p, 42p
micropopulation 51
microseparation 56

Minimum Area Requirement (MAR) 46
Minimum Viable Population (MVP) 3, 23, 33, 45p, 67, 77, 96, 128, 131, 135, 137
model
– mathematical 33, 67pp
– phenomenological 85, 88
– simulation 13
monitoring 113
monogamy 38, 83, 89pp
monophyletic 64
mortality 25p, 82p
– juvenile 82
mortality rate 124, 127p
– extrinsic factor 129
– intrinsic factor 129
mosaic-cycle 2, 43
mountain ridge 110
mutation 39
– rate 35

nature reserve 46
neighbourhood size 37

outbred line 15
outbreeding depression 35
overproduction 38

panmixis 54pp
peat bog 147
persistence 6, 10pp, 17, 40pp
perturbation 7, 45
pitfall trapping 131p
plant
– association 121pp
– eater 122
plasticity 128
polygamy 36pp, 83, 89pp
polygenic control 54
polyphyletic 64
population
– composite 135
– density 40p, 105, 131pp
– density index 133
– dynamics 3p, 10, 68p, 83pp
– effective size 4p, 18, 34pp
– extinct 41p
– fluctuations 29p, 37
– fragmentation 110
– growth 123, 128
– growth rate 124pp
– ideal 34pp
– initial size 124pp
– isolated 115p, 139, 144
– local 23, 43, 105, 119
– local unit 131
– maximum size 6
– minimum size 42, 119, 128
– numbers 124pp
– regulation 129
– single 135
– size 11, 17, 25, 40, 51, 107, 123pp, 131pp
– world 23pp
Population Viability Analysis (PVA) 3
potential
– evolutionary 7p, 18
pregnancy 123p
pressure
– systematic 13, 16
progeny 36, 127

radiation
– adaptive 28, 57
– intralacustrine 56
radio tagging 139
rainforest ecosystem 113
rarity 10, 41
recolonization 82
recovery 73
recruitment 5
regression
– parabolic 124pp
regulation
– density-dependent 68, 86
relaxation 10p, 39
releasing experiments 149p
reproduction 5p, 36, 122p
– rate 84, 88p, 93
– sexual 85p, 91
– success 9, 38
rescue-effect 23, 30
resource
– limiting 77
risk spreading 135
rowan 144pp
r-strategist 2
running activity 131pp

salt concentration 54
sandbar island 109p
– birds 109p
scale 53pp
seabird 25pp
– colony 28
segregation 51pp
selection 7p, 39, 56
– disruptive 56
– pressure 43
separation 63
– microgeographic 54
sex ratio 36p
size class 24
social dysfunction 12

space 39
specialist 105pp
– ecological 107
specialization 56
speciation 51pp
– allopatric 55
– explosive 55, 63
– intralacustrine 55pp
– sympatric 55
species
– composition 113, 135
– endemic 13, 106pp
– loss 10, 113
– migratory 12, 31, 43
– rare 11
spring breeder 132p
spruce
– forest 142pp
– monoculture 147
– plantations 141, 146p
stochasticity 4, 30
– demographic 6p, 35, 69p
– environmental 6p, 68
– genetic 4pp
subpopulation 43, 119, 131, 136
– isolated 137, 149p
– size 150
subspecies 28, 151
successional stage 137pp
survival 67, 124
– mean time 74
– rate 11
– time 74
synchrony 131pp

target species 4
territoriality 128
territory 25p
tetranoids 137pp
Thuringian forest 144pp
transition
– probability 95
trapping series 123p

variability
– genetic 34p, 38
variation
– environmental 44p
– genetic 4, 10, 18, 43pp
– morphological 14p
vulnerability 3p, 30, 81

warbler
– Kirtland's 31
– Seychelles 30p
wetland 25pp

zoogeographic regions 27

Ecological Studies

Volumes published since 1988

Volume 68
Stable Isotopes in Ecological Research (1989)
P.W. Rundel, J.R. Ehleringer,
and K.A. Nagy (Eds.)

Volume 69
Vertebrates in Complex Tropical Systems (1989)
M.L. Harmelin-Vivien and F. Bourlière
(Eds.)

Volume 70
The Northern Forest Border in Canada and Alaska (1989)
J.A. Larsen

Volume 71
Tidal Flat Estuaries: Simulation and Analysis of the Ems Estuary (1988)
J. Baretta and P. Ruardij (Eds.)

Volume 72
Acidic Deposition and Forest Soils (1989)
D. Binkley, C.T. Driscoll, H.L. Allen,
P. Schoeneberger, and D. McAvoy

Volume 73
Toxic Organic Chemicals in Porous Media (1989)
Z. Gerstl, Y. Chen, U. Mingelgrin,
and B. Yaron (Eds.)

Volume 74
Inorganic Contaminants in the Vadose Zone (1989)
B. Bar-Yosef, N.J. Barrow, and
J. Goldshmid (Eds.)

Volume 75
The Grazing Land Ecosystems of the African Sahel (1989)
H.N. Le Houérou

Volume 76
Vascular Plants as Epiphytes: Evolution and Ecophysiology (1989)
U. Lüttge (Ed.)

Volume 77
Air Pollution and Forest Decline: A Study of Spruce (*Picea abies*) on Acid Soils (1989)
E.-D. Schulze, O.L. Lange, and R. Oren
(Eds.)

Volume 78
Agroecology: Researching the Ecological Basis for Sustainable Agriculture (1990)
S.R. Gliessman (Ed.)

Volume 79
Remote Sensing of Biosphere Functioning (1990)
R.J. Hobbs and H.A. Mooney (Eds.)

Volume 80
Plant Biology of the Basin and Range (1990)
B. Osmond, G.M. Hidy, and L. Pitelka
(Eds.)

Volume 81
Nitrogen in Terrestrial Ecosystem: Questions of Productivity, Vegetational Changes, and Ecosystem Stability (1991)
C.O. Tamm

Volume 82
Quantitative Methods in Landscape Ecology: The Analysis and Interpretation of Landscape Heterogeneity (1990)
M.G. Turner and R.H. Gardner (Eds.)

Volume 83
The Rivers of Florida (1990)
R.J. Livingston (Ed.)

Volume 84
Fire in the Tropical Biota: Ecosystem Processes and Global Challenges (1990)
J.G. Goldammer (Ed.)

Volume 85
The Mosaic-Cycle Concept of Ecosystems (1991)
H. Remmert (Ed.)

Volume 86
Ecological Heterogeneity (1991)
J. Kolasa and S.T.A. Pickett (Eds.)

Ecology of Equids and Their Impact on the Camargue (1992)
P. Duncan

Volume 88
Pinnipeds and El Niño: Responses to Environmental Stress (1992)
F. Trillmich and K.A. Ono (Eds.)

Volume 89
Plantago: A Multidisciplinary Study (1992)
P.J.C. Kuiper and M. Bos (Eds.)

Volume 90
Biogeochemistry of a Subalpine Ecosystem: Loch Vale Watershed (1992)
J. Baron (Ed.)

Volume 91
Atmospheric Deposition and Forest Nutrient Cycling (1992)
D.W. Johnson and S.E. Lindberg (Eds.)

Volume 92
Landscape Boundaries: Consequences for Biotic Diversity and Ecological Flows (1992)
A.J. Hansen and F. di. Castri (Eds.)

Volume 93
Fire in South African Mountain Fynbos: Ecosystem, Community, and Species Response at Swartboskloof (1992)
B.W. van Wilgen et al. (Eds.)

Volume 94
The Ecology of Aquatic Hyphomycetes (1992)
F. Bärlocher (Ed.)

Volume 95
Palms in Forest Ecosystems of Amazonia (1992)
F. Kahn and J.-J. DeGranville

Volume 96
Ecology and Decline of Red Spruce in the Eastern united States (1992)
C. Eagar and M.B. Adams (Eds.)

Volume 97
The Response of Western Forests to Air Pollution (1992)
R.K. Olson, D. Binkley, and M. Böhm (Eds.)

Volume 98
Plankton Regulation Dynamics (1993)
N. Walz (Ed.)

Volume 99
Biodiversity and Ecosystem Function (1993)
E.-D. Schulze and H.A. Mooney (Eds.)

Volume 100
Ecophysiology of Photosynthesis (1994)
E.-D. Schulze and M.M. Caldwell (Eds.)

Volume 101
Effects of Land Use Change on Atmospheric CO_2 Concentrations: South and South East Asia as a Case Study (1993)
V.H. Dale (Ed.)

Volume 102
Coral Reef Ecology (1993)
Y.I. Sorokin (Ed.)

Volume 103
Rocky Shores: Exploitation in Chile and South Africa (1993)
W.R. Siegfried (Ed.)

Volume 104
Long-Term Experiment With Acid Rain in Norwegian Forest Ecosystems (1993)
G. Abrahamsen et al. (Eds.)

Volume 105
Microbial Ecology of Lake Plußsee (1993)
J. Overbeck and R.J. chrost (Eds.)

Springer-Verlag and the Environment

We at Springer-Verlag firmly believe that an international science publisher has a special obligation to the environment, and our corporate policies consistently reflect this conviction.

We also expect our business partners – paper mills, printers, packaging manufacturers, etc. – to commit themselves to using environmentally friendly materials and production processes.

The paper in this book is made from low- or no-chlorine pulp and is acid free, in conformance with international standards for paper permanency.

Printing: Mercedesdruck, Berlin
Binding: Buchbinderei Lüderitz & Bauer, Berlin